Praise for
Geek Dad:

"Includes scores of illustrated projects that a parent and child can do together, using many materials that barely existed a few years ago . . . Stand back and watch the fun ensue." —*The New York Times*

"These are truly fun, inspired, and even *educational* projects you can do with your kids." —*Wired*

"There are projects that require computer skills or some knowledge of electronics; mostly, though, they require inquisitiveness and imagination . . . The bottom line—just as it was with our fathers and grandfathers—is doing something with your kids that is fun and interesting for all parties." —*Chicago Tribune*

"Call it *Revenge of the Nerds V.* . . . A book that embraces geek culture." —*San Francisco Chronicle*

"A how-to guide that includes nighttime kite flying, electronic origami, and cyborg jack-o-lanterns." —*USA Today* gift guide

"Craft projects to be shared by kids and techie dads are collected by the author of the *GeekDad* blog on Wired.com. The best ideas involve the making of a comic strip out of Legos and flying a light-rigged kite at night." —*Milwaukee Journal Sentinel*

"Projects for science enthusiasts of all ages." —NPR's *Science Friday*

"A great book . . . This is fun you can have with your kids, with your coworkers, with your spouse, and your friends." —arstechnica.com

"[*Geek Dad* features] dozens of geek-friendly activities and rainy-day projects for parents (not just dads, we should note) and kids to enjoy, from creating your own comic strips to building a working lamp out of CDs and LEGO bricks. Pick it up in time for Father's Day, and be sure to let Dad know he has to share." —babble.com

"Read this crafty book for ideas to share your love of science, technology, gadgetry, and MacGyver. . . . Soon, together, you can rule the galaxy as father and son. Mwahaha."　　　　　　　　　—boston.com

"The book provides an easy gateway to spending purposeful time with your kids and sharing experiences that they'll never forget, all in the spirit of tech-savvy DIY. Now that's an idea to geek out about."
　　　　　　　　　—gearpatrol.com

"I think many geek dads (and okay, geek moms too) will appreciate the hands-on approach that fills the void that kids just don't get today in school, especially in the American culture of 'teach to the test.'"
　　　　　　　　　—geekcowboy.net

"Learn to create all kinds of great geeky stuff and fun with your offspring, including the ultimate in summer fun—a super 'slip and slide.'"
　　　　　　　　　—examiner.com

"What a fun, educational but not in a boring way, book!"
　　　　　　　　　—meghansmusings.wordpress.com

"Full of all kinds of cool projects you can do with your kids, the book is a must-have for any dad (hint: perfect father's day gift!) who loves to get geeky with the little ones."　　　　　　　—mentalfloss.com

"A compendium of fun and geeky projects for kids to do with their parents."　　　　　　　　　—MAKEzine.com

"An easy-to-read, adroitly written craft book."　　　—neatorama.com

"I would encourage any adult with latent geek-like tendencies to acquire this tome to realize their full potential."　　　—STliving.com

"A perfect manual for sharing your geekiness with the next generation."
　　　　　　　　　—thatsbadass.com

"Denmead's subcategories for each project—concept, cost, difficulty, duration, and reusability—help readers easily navigate the book to find items they want to make that match their budgets, skill-levels, etc."

—urbanbaby.com

Praise for
The Geek Dad's Guide to Weekend Fun:

"Packed with delightfully fun projects for a rainy day or romping in the sun. The ratio of building and playing is just right."

—Chris Anderson, editor in chief, *Wired*

"The Geek Dad blogger on Wired.com is also an engineer, so none of his enticing roster of weekend projects is likely to collapse—or explode. Each DIY project comes with measures of cost, difficulty, duration, and the tools and materials you'll need."

—*The Globe and Mail*

"Not only is it loaded with a new batch of great things to do with your kids . . . there's also a selection of projects and prose by some very cool geeky celebrities, like Ken Jennings (the *Jeopardy!* champ), Chris Anderson (our founding geek-father), Rod Roddenberry (keeper of the Star Trek flame), Patrick Norton (tech guru), David Hewlett (sci-fi acting fave), and more."

—Wired.com

"Local dad, tinkerer, and *Wired* blogger Ken Denmead's new book, *The Geek Dad's Guide to Weekend Fun,* shows you how to collaborate with your budding scientist on everything from homemade robots and dry-ice ice cream to chocolate Matchbox cars."

—DailyCandy.com

"All of the projects have a practical, teachable component to them but kids don't have to know it—just knowing that a trebuchet is a medieval weapon will guarantee their participation."

—Columbian.com

THE
GEEK
DAD

Book for Aspiring
Mad Scientists

The Coolest Experiments for
Science Fairs and Family Fun

KEN DENMEAD

GOTHAM
BOOKS

GOTHAM BOOKS

Published by Penguin Group (USA) Inc.
375 Hudson Street, New York, New York 10014, U.S.A.
Penguin Group (Canada), 90 Eglinton Avenue East, Suite 700, Toronto, Ontario M4P 2Y3,
Canada (a division of Pearson Penguin Canada Inc.); Penguin Books Ltd, 80 Strand, London
WC2R 0RL, England; Penguin Ireland, 25 St Stephen's Green, Dublin 2, Ireland (a division of
Penguin Books Ltd); Penguin Group (Australia), 250 Camberwell Road, Camberwell,
Victoria 3124, Australia (a division of Pearson Australia Group Pty Ltd); Penguin Books
India Pvt Ltd, 11 Community Centre, Panchsheel Park, New Delhi—110 017, India; Penguin
Group (NZ), 67 Apollo Drive, Rosedale, Auckland 0632, New Zealand (a division of Pearson
New Zealand Ltd); Penguin Books (South Africa) (Pty) Ltd, 24 Sturdee Avenue, Rosebank,
Johannesburg 2196, South Africa

Penguin Books Ltd, Registered Offices: 80 Strand, London WC2R 0RL, England

Published by Gotham Books, a member of Penguin Group (USA) Inc.

First printing, November 2011
10 9 8 7 6 5 4 3 2 1

Copyright © 2011 by Ken Denmead
All rights reserved
Illustrations by Bradley L. Hill

Gotham Books and the skyscraper logo are trademarks of Penguin Group (USA) Inc.

LIBRARY OF CONGRESS CATALOGING-IN-PUBLICATION DATA
has been applied for.

ISBN 978-1-592-40688-3

Printed in the United States of America
Set in Apollo MT Std
Designed by Sabrina Bowers

While the author has made every effort to provide accurate telephone numbers and
Internet addresses at the time of publication, neither the publisher nor the author
assumes any responsibility for errors, or for changes that occur after publication. Further,
the publisher does not have any control over and does not assume any responsibility for
author or third-party websites or their content.

Contents

Special Thanks

As always (and it amazes me I can actually say "as always" here), my most special thanks go to my wife, Robin, for her partnership in life and helping make this all possible, since if it weren't for her, I wouldn't be a dad. Also, she's a teacher, and I've learned a huge amount about education via osmosis from her and through her passion for her students. What she does on a daily basis deserves greater accolades than I can provide.

Thanks to my boys, Eli and Quinn, for being patient with their dad when he's working his second and third jobs (the blogging and the book-writing), and still being excited about playing golf or *Portal 2* with me.

And special love to all our pets, who don't really know what's going on, but love us as much as we love them. Brody (dog), Sassy (dog), Short Round (dog), and Barkley (fish).

I also think for this book that it's important to thank the science teachers who helped feed the science-geek side of my personality. Especially Mr. Albee (aka "boom-boom") in Chemistry and Mr. Barrows in AP Physics at the Head-Royce School.

And my ongoing appreciation goes out to all the people working hard to make science fun and popular and keep kids interested in the things that make our world run. In no particular order: Bill Nye, The Mythbusters (Adam Savage, Jamie Hyneman, Kari Byron, Grant Imahara, and Tory Belleci), Professor Michio Kaku, Dr. Kirsten "Kiki" Sanford, Phil Plait, Gever Tulley, Bill Gurstelle, and more. Keep the fire alive!

Introduction

"We are dreamers, shapers, singers and makers. We study the mysteries of laser and circuit, crystal and scanner. Holographic demons and invocations of equations."

—ELRIC THE TECHNOMAGE (*BABYLON 5*, SEASON 2, EPISODE 3 —PLAYED BY MICHAEL ANSARA)

We seem to have gotten further and further away from an honest understanding and use of science in our daily lives. When emotional and political motivations color the use of scientific results, the true purpose of science gets lost. Take, for example, the recent controversy over issues like vaccination. Vaccination works. The overwhelming body of scientific evidence proves that while there are individual cases of poor reactions to vaccinations, the health of our society as a whole is significantly improved as a result of them. Yet the science surrounding them has been twisted in recent years, alarming parents and creating doubt. The result is more risk to children who are not vaccinated.

Science is nothing short of the search for truth. Science has no bias. Science has no spin. Science merely gives us the unvarnished answers to the questions we ask. Whether we ask the right questions, or understand the answers, is totally our problem. But it also affects our kids because they learn how to question the world around them by watching us. If we are inquisitive, they'll likely be inquisitive. If we are skeptical and demand proof before accepting dubious claims, they'll probably be skeptical as well, and smarter for it.

That is why it's vital for us to be involved in our kids' science education. In fact, by learning the rigorous logical methodology of science, kids can discover how to approach other school subjects and even other problems they encounter through their lives. Understanding how to state a hypothesis, derive the tests that must be met to prove it, effectively analyze the results, and provide a conclusion is central to science *and* the key to so many other pursuits.

Of course, as with anything, a big part of getting kids engaged with science is getting them excited, and luckily, beyond all the skills science can teach about methodology, logic, and the rigorous application of principles, it can be really fun, too. Digging into science is like going to Hogwarts and learning the secret magic that governs every aspect of day-to-day life. And when you master science, you can learn how to use and adapt its principles for your own purposes and even entertainment.

For example, I have a fondness for Avogadro's number (6.022 × 10^23). Huh? What's that, you ask? It's a funny little number that tells you how much of an elementary particle—an atom or molecule—you have when you have a mole of it. What's a mole? Well, the easiest way to understand it is that a mole of a substance is the mass in grams of the substance equal to its molar mass (which you can usually look up for each and every element on a periodic table). For example, take water. H_2O. In round numbers, the molar mass of water is 2 × 1 (for hydrogen) + 1 × 16 (for oxygen) = 18 grams per mole. Therefore, in an 18 gram sample of water, there are 6.022 × 10^23 molecules of water. From this number you can find the density of water, or even the mass of one molecule. Handy, if obscure, science, eh?

Which isn't to say that science education is about learning trivia. Avogadro's number is actually a vital part of chemistry and physics. The point is, when you know the trivia of science, you know how things work. I mean *how things work*! And that's awesome.

Which brings us to the purpose and goal(s) of this book. Even more than the other *Geek Dad* books, this book is about enjoying science with your kids. Every kid is going to have to deal with science at some point in their education, and probably have to do something for a science fair. I'm hoping this book will give every family a bit of help in getting into the "playing with science" mind-set (key point: Have fun with it!), a better understanding of scientific methodology, and a good selection of basic ideas for potential science fair projects.

What this book isn't is the be-all and end-all, comprehensive list for all the possible science fair projects out there. These are suggestions. You can do these experiments at home together, too, for fun and learning (and they are fun!). But if you want to help your kids do any of them as science fair projects, make sure you don't tell them the answers! The whole point of science fair projects is that while kids may suspect the answers, they should be going through a journey of discovery as they do the experiments. That's why, on some projects, you'll see "<SPOILER>" warnings—pieces of information to withhold from them if you're going over the book as a resource for their class work.

Of course, if you're just doing it all for entertainment alone, no worries. Have at it! Enjoy science and the amazing things you can explore in your own home. And celebrate learning a little something magical every time you do.

WHAT'S INCLUDED IN THIS BOOK

At its most basic level, the Scientific Method is about approaching any task in a logical and rigorous manner. We ask ourselves a question like "How far can I throw a rock?" Then we hypothesize an outcome based upon what we know and decide how we're going to determine if the hypothesis is correct by setting up an experiment, defining parameters, and ensuring that nothing can corrupt the results. Then we perform the experiment, gather data, study it to see if it proves or disproves our hypothesis. If need be, we can then refine our hypothesis and repeat. It sounds dry, but it's also very powerful. It's how we gather irrefutable answers about the way our world works.

The projects included in this book cover a broad swath of scientific principles and approach those principles with an eye toward having fun while learning about the science and the tenets of the Scientific Method, as well as some existing bits of the history of science and how we know what we know. While science fairs can be serious endeavors, it never hurts to approach your project with a sense of whimsy, which is why the projects are organized the way they are, into four rather geeky groups:

Experiments for Moonbase Alpha: The experiments in this section imagine a future when we can cast off the shackles of gravity and begin the colonization of our closest neighbors in the solar system. If this happens, any aspiring mad scientist will need to be prepared for life in space and know how to sustain their minions of moon people. The projects in this section look at both the hard and soft sciences needed for life aboard a space station.

Inside the Mad Scientist's Kitchen: Molecular gastronomy is all the rage these days, but what does it mean? It's all about using science in the kitchen to do things with food that haven't been tried before. But the best chefs and mad scientists have always known that food and science are inseparable. The projects in this section all have to do with playing with your food in strange and exciting ways.

Apocalypse Survival Science: There's always the chance that an apocalypse awaits us in the future. Whether it's zombies, robots, a horrible virus, the next ice age, or a dozen other cheery armageddons, science fiction often revolves around a good yarn about the decimation of the planet, leaving a few desperate survivors to find their way in a world bereft of technology and other niceties. Such a scenario would require relearning basic scientific principles and how they can apply to life in a harsh environment. The projects in this section will help prepare your little mad scientist to lead the remnants of society into a new dawn.

Fun with Fire and Electricity: As our heroes on *Mythbusters* (probably the best show on television that celebrates how cool science can be) have shown us over the years, playing with fire can be fun, and electricity is electrifying. The projects in this section all explore the powerful, creative, and destructive energies we find in the natural world.

NEXT, ON A VERY SPECIAL EPISODE OF *GEEK DAD*: FIRE/ELECTRICAL/CHEMICAL SAFETY FOR THE ASPIRING MAD SCIENTIST

While the goal of this book is to approach science and science fair projects with a bit of fun and whimsy, any experimentation has its risks, and every scientist must respect those risks and prepare for them. Why do you think scientists (and doctors and pharmacists) wear those white lab coats that hang to the knees and have long sleeves? No, it's not just a fashion statement or product placement for the bleach industry. It's about safety. If you have a clean white lab coat, it's much easier to see when you might have splashed a potentially dangerous chemical on you.

So, let us take a sober moment and talk about keeping you and your evil minions safe (mad scientist liability insurance is a real budget-buster, after all).

Here are some things you're going to want to have when performing the more rigorous of the experiments included in this book:

▶ **Lab Coat:** If you don't want to lay out the extra money for new duds, kitchen aprons and long-sleeve shirts can be a reasonable substitute. On the other hand, you can find lab coats for under $20 online pretty easily, and then you'll really look the part!

▶ **Eye Protection:** There may be splashes or smoke involved in some of these projects, and you MUST protect your eyesight. Even if a patch or hideous scar is often a great way to look extra menacing as a mad scientist, having one or both actually suggests you're a sloppy mad scientist, and your nemeses will respect you less. Get safety glasses (cheap and easy to find at your local hardware store), or even better, proper mad scientist goggles (if you're a proper Dr. Horrible fan, search for the Hobart 770129 Oxy/Acet Goggle on amazon.com—you'll be very happy).

▶ **Hand Protection:** When handling chemicals, standard latex gloves should be fine, but because there will be fire in some of these projects, you may want to step up your game a little. The challenge with fire-resistant gloves is that they tend to be bulky, and significantly hamper dexterity. A good compromise is a pair of the heat-resistant oven gloves you can find online or in many kitchen-supply stores. Some even come with silicon finger pads to increase grip capacity. The real thing to be careful about with these is that you don't start using them all the time and then one day, forget you took them off and grab something hot.

▶ **Fire Extinguisher:** It's absolutely vital that you have at least one, if not more than one, fire extinguisher at hand and readily available when working with fire for these experiments. Most big-box stores (Costco, Walmart, Target) carry inexpensive multi-packs of fire extinguishers for home use. Get them, check them every month and, every time you start a project, make sure they're still properly filled and pressurized. A backup is to have a box of baking soda or salt handy to throw on a fire and smother it (just make sure you don't use flour, as we'll show you!).

▶ **Basic Common Sense**: Sometimes, even when you have the best intentions, this is the hardest thing to keep at hand. All of the experiments in this book are meant to be done by parents with their kids. If the kids are doing them as science fair projects and parents can't directly perform the work, PARENTS MUST STILL ENSURE ALL SAFETY MEASURES ARE BEING OBSERVED. If you are doing these experiments together, you must lead by example and take care with every step. And especially, remember:

- Never leave your budding mad scientist alone with an open flame or active chemical reaction. If there's something really pressing (phone call, person at the door, UFO sighting), just put the fire/reaction out first, even if it's premature. You can always start over again when you get back.

- Keep everything neat and organized. Science is a process, and sometimes a messy process. Again, being deliberate, slow, and organized is the best way to get the cleanest results, and to maintain a safe work area. Keep your "lab" clean and uncluttered, and make sure everything is labeled or clearly identified so nobody grabs the wrong material in a moment of confusion. Indeed, there should never be any confusion in the lab—even in a mad scientist's lab!

- Keep things fun, but calm and controlled. Doing science is a blast, but for the most part, we don't want to cause a blast (at least not one we didn't know was going to happen). While everyone involved can be having fun, horseplay (why don't we call it people-play?) in the workspace must be discouraged. If you are managing multiple aspiring mad scientists alone, enforce upon them the importance of maintaining a proper scientific atmosphere.

EXPERIMENT INFORMATION

At the start of each experiment, you'll see a table with summary information to give you an idea of what to expect from it, as well as some symbols not unlike what you see in a restaurant or hotel review to explain cost and difficulty. Here's a legend to explain their meaning.

EXPERIMENT	TITLE OF THE EXPERIMENT
CONCEPT	A quick overview of the project so you can decide if it's of interest to you.
COST	$ — $0 to $25, $$ — $25 to $50, $$$ — $50 to $100, $$$$ — $100 on up. In many cases I'll exclude the cost of tools and materials that are so common you probably already own them or could easily find or borrow them.
DIFFICULTY	⚙ — easy for primary school–age kids to grasp and enjoy, ⚙ ⚙ — for secondary school–age and up, ⚙ ⚙ ⚙ — for junior high and up, ⚙ ⚙ ⚙ ⚙ — high school age. There is a wide variety to these experiments, from the very simple to the rather complex. Some of them can even be adjusted to be more or less in-depth. It's up to you to gauge your kids' ability level and attention spans, and pick the right experiments to share with them. After all, you know your kids better than I do!
DURATION	☀ — 0 to 1 hour, ☀ ☀ — 1 hour to 1 day, ☀ ☀ ☀ — 1 day to 1 week, ☀ ☀ ☀ ☀ — 1 week or longer. Please note that this duration scale has changed from previous *Geek Dad* books. Scientific experiments often take significant time for data gathering, and this expanded scale reflects that.
DEMONSTRATION OR EXPERIMENT	Explains whether the experiment will serve as a demonstration of a scientific principle, or allow for further development, perhaps for a science fair.
TOOLS & MATERIALS	A list of the basics required to perform the experiment. I'll do my best to suggest how to make do without buying too much.

As with the previous *Geek Dad* books, I've tried hard to include projects/experiments that don't cost a lot of money to put together, and that come in a range of difficulties. Unlike the previous books, though, you'll see more projects that have a longer duration time.

While many of these experiments (or at least the demonstration versions of them) can be done in a couple of hours, the more involved ones, especially those intended for use in science fairs, may take multiple days, weeks, or even months. Before you and your child decide to use one of these projects for a kids' science fair, consider how much time there is to do the experiment and prepare the presentation for it.

And I hope you'll enjoy the tongue-in-cheek references to mad scientists in movies and literature. While such mad scientists are often bad characters who believe the ends justify the means, they also often use science in a playful way, and that's what I want your family's takeaway to be here. Doing science together can give your family hours and hours of entertainment, and it'll teach you a lot about many other things along the way. Sharing this kind of learning with our kids is one of the most valuable things we can do as parents.

So, to quote a mad scientist and his henchman, famously from *Animaniacs*:

Pinky: "What are we going to do today, Brain?"

Brain: "Same thing we do every day, Pinky: Try to take over the world!"

EXPERIMENTS FOR MOONBASE ALPHA

Extracting Your Own DNA

Idea by Kathy Ceceri

O
kay, so you've set up your orbiting base. You have robot minions at your beck and call, and your plans for world domination are just pouring out of your fevered mind. The problem
is, while robot minions are all well and good for taking care of the
heavy lifting (for, say, pushing prospective interlopers out the airlock), they're not self-starters. They don't think independently. And
while they may be good multi-taskers, you're the one who has to assign them each of those tasks, with detailed instructions that are as
clear as possible. Otherwise, you're likely not to get the results you
want. (Imagine the mishap that could happen if you commanded
your robots to "nuke some dinner." To get the help you really need,
you'd end up spending all your time managing your troops and
none of it "mad-scientisting.") The best solution is to find someone
who is as smart as you to direct the robots. Indeed, you need another you.

Time for some cloning! And what's the most important first step
toward proper cloning? Why, extracting one's own DNA, of course.

EXPERIMENT	EXTRACTING YOUR OWN DNA
CONCEPT	Use household chemicals to collect the DNA from the nucleus of cells in your cheeks.
COST	$
DIFFICULTY	⚙ ⚙
DURATION	☼
DEMONSTRATION OR EXPERIMENT	This project works better as a demonstration.
TOOLS & MATERIALS	• Isopropyl (rubbing) alcohol (70% or 91%) • Clear plastic cups • Plastic spoons • Small plate (preferably disposable) • Drinking water • Salt • Dish soap • Blue food coloring (optional)

A Little Bit of Science History

Deoxyribonucleic acid, aka DNA, was first identified in its basic parts (nucleic acids) in 1869 by Swiss physician and biologist Johannes Friedrich Miescher, who named them *nuclein* because they came from the nucleus of white blood cells. We know now that every form of life (as we know it here on Earth) uses nucleic acids (either DNA or its cousin RNA—ribonucleic acid) as the means by which the instructions on how to replicate and propagate themselves at a cellular level are stored.

Oswald Avery determined in 1944 that DNA carries the genetic information that is transferred to newly created cells. Each strand of human DNA is divided into 23 pairs of chromosomes, which in turn contain hundreds or thousands of genes. Genes record information in the form of chemical codes about how to build the proteins and other molecules that make up living organisms.

DNA is one of the longest type of polymers, or chains of molecules. Strands of human DNA are 6 feet long. In 1950, scientist Rosalind Franklin used x-ray crystallography to find that DNA is made up of two long polymers, or chains of molecules, twisted into a shape called a double helix. (And indeed, if the politics of publishing in scientific journals had worked differently, her work might have won *her* a Nobel Prize, instead of James Watson and Francis Crick, who built off her work to discover that the strands were connected by crossbars, like a ladder. They won the Nobel Prize in 1962 for describing the complete picture of DNA's structure.)

This is a pretty easy experiment to perform, and the idea behind it is totally cool. What we're going to do is literally rip some DNA out of your body and stretch it out to the point where we can see it without the aid of any special tools (though if you want to take a peek with a microscope, or even some pictures of it, you're welcome to). But first:

EXTRACTING YOUR OWN DNA

STEP 1: You need to chill the alcohol, but not freeze it. Put your bottle of alcohol in the freezer while performing the next steps, but monitor it regularly to make sure it doesn't start to freeze. If you see crystals forming, take it out, shake it up, and put it in the refrigerator for a spell before returning it to the freezer.

STEP 2: We each have literally trillions (that's 1×10^{13}) of cells in our bodies that include genetic information (and that doesn't even count all the microbes and other bugs we carry around in and on us in symbiotic relationships). You're going to gather just a few cells (don't worry, you won't miss them) from the inside of your cheeks by swishing a special solution around in your mouth.

In the plastic cup, mix $\frac{1}{4}$ teaspoon of salt in $\frac{1}{4}$ cup of water. Swish the saltwater around in your mouth for a minute, making sure it reaches the inside of your cheeks. Spit all the water back in the cup. Don't swallow, or you'll lose your samples. Plus, it doesn't taste all that good. (As an alternate, you can use plain Gatorade, which tastes better. But come on, this is science! Science isn't supposed to taste good!)

What's the Science Here?

Cells from the inside of your cheek mix with the saltwater and are carried away when you spit (along with your saliva, and possibly some remnants of whatever you had for lunch). Page 18 has a photograph of cheek cells under a microscope. If you looked at the sample under a microscope, you'd be able to see the nucleus inside each cell where the DNA is all clumped up waiting for us to let it out.

STEP 3: Now comes the jailbreak. You're going to release the DNA from your cheek cells. Put a drop of dish soap on the plate. Touch the bottom bowl of the spoon to the soap so you get just the hint of a drop on it, and then dip it in the cup of saltwater. Gently stir once or twice.

The cheek cells are contained within membranes that are made up of fats. The soap solution breaks down the fat molecules, just like soap breaks down grease on your dishes, and releases the contents of the cell.

STEP 4: Go get your not-quite-frozen alcohol from the freezer. Pour about ¼ cup into a second plastic cup. You can also add a drop of blue food coloring to make the alcohol easier to see. Stir until evenly mixed. Take the saltwater cup and tilt it to one side. Hold the cup of alcohol up so that the lip touches the tilted cup. *Slowly* pour a little alcohol down the inside of the cup so that it floats on top of the saltwater without mixing. This process will layer, or stack, your drink—the different specific gravities (densities) of the liquids will allow you to put them together in a glass but not let them mix. Continue until there is about ⅛ of an inch of alcohol on top of the saltwater.

STEP 5: Now you wait—but not too long! Within a couple of minutes, you will see long strands or clumps of a sticky white substance start to come together in the alcohol layer. This is your DNA, released from the thousands of cheek cells in the saltwater. Because the protein-based DNA cannot dissolve in the chilled alcohol, it precipitates out of the saltwater solution and appears as a solid.

And that's it, you have your DNA! Interestingly enough, this is pretty much the methodology used when scientists (you know, the normal, boring "not-mad" kind) need to extract DNA to study. If you look really, really close you might even see the double helix!

Okay, I lied about that last part. You can't see the helix with the naked eye, or even with a super-powered electron-scanning microscope. But you can remove the DNA to look at it through a normal at-home compound microscope by dipping a toothpick into the alcohol layer and twirling it to gather up the sticky DNA. Place the sticky stuff on a glass microscope slide and take a peek—you won't be seeing down to the molecular level, though, just big clumps of the precipitated genetic material. You can also save the DNA in a small, clear container with a little extra saltwater, just in case you need it for that home-cloning project down the line.

What Else Can I Do?

To make the demonstration more interesting, or perhaps as part of a proper hypothetical experiment, you could try to gather samples of cells from other organisms and look at their DNA. Do you have a dog, cat, or other mammalian pet? While you can't easily get it to swish the saltwater compound in its mouth, you could try getting a Q-tip damp with saltwater and swabbing the inside of their cheeks to gather cells (assuming your pets will allow this—you might need them to sign a waiver). If you can, do this with a couple of Q-tips and stir each one in the cup of saltwater. Then perform the rest of the experiment as described above and see if you can get animal DNA to precipitate out. You can try to answer questions like "Does animal DNA look the same as human DNA?" And, of course, it's a good idea to know how to clone your pet since you can never have too much unconditional love (though you may need to get your robot-lackeys to take care of the feeding and walks while you attend to world dominance).

Space Agriculture

Idea by Dave Banks

As mankind inevitably reaches out into space to search out new worlds and new civilizations (to conquer, MUAHAHAHA!), one of the most important things we'll have to worry about is the food we'll have on the long intra- and inter-solar journeys. Freeze-dried foods and nutritional supplements do the job, but wouldn't it be great if we could invent a way to enjoy fresh fruits and veggies during our journey to the stars?

To shed some light on how we'd successfully grow plants on our long voyages, we need to know how plants will grow in space. That means thinking about light and gravity. What are the impacts of gravity and directed growing light on the staple plants we want to take with us? This experiment will reveal those secrets.

EXPERIMENT	SPACE AGRICULTURE
CONCEPT	All living creatures are affected by external stimuli like sight, sound, or touch. But sometimes what affects our behavior may not be something we notice directly via our senses. These two experiments show how—over time—a living creature's development can be radically affected by external stimuli; effects called *tropisms*.
COST	$
DIFFICULTY	⚙ ⚙
DURATION	☀ ☀ ☀ ☀
DEMONSTRATION OR EXPERIMENT	Can be a demonstration, but can also be a good science fair experiment.
TOOLS & MATERIALS	• 2 1-gallon milk cartons (the cardboard kind) • Potting soil • Sealable plastic sandwich bag • Paper towels • Foil or heavy tape • Tray or plate to catch drainage • Bean seeds (lima, navy, or radish)

If you go for a drive along a coastal road where the wind is always blowing in from the sea, you'll notice that the trees all look like cartoon characters who just survived an explosion—they have been blown in one direction for so long that they look frozen in an awkward position This is a tropism, meaning an effect of and external stimulus on a biological organism. In this case, the near-constant force of the wind has trained the trees to grow in the path of least resistance. If you remember the mathematician character in *Jurassic Park* pointing out (over and over again) that "life finds a way," tropisms are an excellent example of that truism.

Life, including plants that don't have consciousness per se, can react to stimuli and adapt physically by altering their growth patterns. But how? What happens when an external stimulus is introduced, removed or changed, and influences an organism's vital needs? We'll find out the answer through the two experiments below.

There are many examples of tropism—touch, water, and chemicals can all act as tropism agents—but in these experiments, we'll look at two very basic examples that will help simulate the unique conditions you'll find on your deep space voyage. One is phototropism, the effect of light on growth; and the other is geotropism, the effect of gravity on growth.

PHOTOTROPISM EXPERIMENT

If you're performing this experiment as part of a science fair, first consider your hypothesis. How do you think messing about with the light source of your plants will affect their grown patterns, if at all?

STEP 1: Take two 1-gallon cardboard milk cartons (cleaned out!) and punch about a dozen holes in the bottoms for drainage.

STEP 2: Wrap the sides of both cartons in foil or heavy tape to make sure no light reaches inside (even light filtering through the cardboard).

STEP 3: Fill bottom of each carton with some potting soil, about 2 inches deep and lightly tamped down.

STEP 4: Place several lima bean seeds, evenly spaced in soil, and water until it starts to drain out of the holes in the bottom (sit the cartons in or over a plate or tray to catch the water run-off).

STEP 5: In one carton (write "Control" on the outside), close the top to leave the bean seeds in the dark. In the other carton (write "Experiment" on the outside), cut a 1-inch by 1-inch square on one side of the carton about 3 inches down from the top of the carton.

STEP 6: Close the tops of both cartons and seal them so that no light can get in through the top (the only light available to any of the seeds should be the hole you cut in the "Experiment" carton).

STEP 7: Place near a window (again, with a tray or plate beneath cartons to catch drainage!). You can open the stops to water and observe each carton on a daily basis, but make sure to close them when you're done. After a week, you should see enough growth to make a scientific observation.

Your project data should answer the following questions: Did the control seeds grow? Which beans are taller? When you look inside the "Experiment" carton, are bean stems leaning toward the hole? You can use a ruler to measure vertical and horizontal growth, to see how far the bean stems lean away from being fully vertical. Track growth rates for each carton to see how much of a difference the light actually makes to plant growth.

It can also be very helpful to take pictures of each of the carton interiors when you record your data. Set up a tripod or other support device for your camera so that you can take a picture looking down into each carton from exactly the same height and orientation each time. This way, when you compare photographs, the perspective will be the same. These visuals will look great on a science fair report board.

What Else Can I Do?

Okay, that's the basic phototropism experiment, but you can take this so much further. If you want a more in-depth experiment, try these variations:

- ► Do this experiment with one control and four test containers, each with their sunlight hole facing in a different direction, and then compare them—is the phototropism effect more obvious?
- ► Try a variety of hole sizes or orientations to see if there's any difference in the growth.
- ► Try two controls—a completely dark one, and one with the top left open all the time—to see the difference between top light and side light.
- ► If you have holes on two adjacent sides, does the effect average out, causing the plants to grow toward the corner?

Another experiment is the "Maze" variant:

STEP 1: Place a fast-growing plant (like an ivy) in the bottom of a large cardboard box that is set to open from the side, like a door, so you can access the plant to water it (make sure no light enters from around the edges of the door).

STEP 2: Above the plant, tape a partition into the box that cuts off about ¾ of the box on the left, a little above the plant. Then, do the same again a few inches above the first partition, leaving space on the right open. You're creating a little maze.

STEP 3: Above the upper partition, cut out a large hole and place the box near a window, with the hole side facing toward the light. Tend to the plant as needed and watch as the plant grows around the maze to reach toward sunlight.

GEOTROPISM EXPERIMENT

Astronauts have done many experiments growing plants in zero gravity and the results always seem to be that roots need to know which way is down to grow properly. While we don't have the capacity yet to create a zero-gravity testing area (though we'll certainly make figuring out how to do that a mad scientist priority!), we can play with directional changes in gravity pretty easily.

STEP 1: Take a couple of paper towels and get them wet. Place them along the bottom of a sandwich-size sealable plastic bag.

STEP 2: Place a few bean seeds on top of a paper towel and plastic bag (with air inside!). Make sure beans are between paper towel and wall of plastic bag. Using Magic Marker, write "Control" on the side of this bag.

STEP 3: Make another, similar bag as described in steps 1 and 2, but label this one "Experiment." Do your best to make sure the amount of water, air, and number of seeds are exactly the same.

STEP 4: Tape the bags to a wall, chair, or other surface where they can get sunlight. The bags can even be tacked to the outside of your house, taped to a swing set, hung from a clothesline, or whatever.

STEP 5: Check the bags daily for growth, and record observations. Moisten (but do not soak) the paper towels as needed to keep the growing medium viable, making sure to keep things as identical as possible in both bags.

STEP 6: Once beans sprout (in 3–5 days), rotate the "Experiment" bag 90 degrees and re-hang in the same place but in this new orientation. Leave the "Control" bag as it was before. Continue moistening the towels and observing.

STEP 7: After several days, did the "Experiment" sprouts change their growth and bend toward the skies again? What did the roots do? Rotate the experiment bag again and observe. Always leave the "Control" bag in its initial orientation so you have a baseline to compare to.

Data recording for this experiment is far more observational, unless you want to use a ruler or calipers and maybe a protractor to take detailed measurements of the angle of growth away from vertical.

Again, it's always good to take pictures, and this project could be an awesome one to consider making a stop-motion movie with. If you can set up your experiment bag on a wall and put a camera on a tripod in front of it, you can try taking pictures every hour or two when you're at home (there are also apps for some camera-equipped smartphones to allow setting timers for taking stop-motion pictures, so you can leave it during the day and get regular snapshots). Being able to show off a movie with your report that demonstrates the growth pattern would be great!

What Else Can I Do?

Another excellent space agriculture study you might wish to try is the hydroponics project from the original *Geek Dad* book (the red one). Hydroponics are likely to be the method of choice for growing plants in space. Studying different growth media, watering, or oxygenation techniques are excellent experiments as well.

DIY Mind Control

For many mad scientists, world domination is not only a dream, it's a responsibility. One is burdened with the weight of over-whelming intelligence and insight into humanity, combined with the technical skill to pull off complicated schemes that lesser minds cannot even fathom.

Of course, you're just doing what's right for the greater good! What's a surefire way to help you achieve your goals if you are able to use its power wisely? (and you must promise to use it wisely and responsibly). Mind control! This experiment will show you how to subtly affect other people's minds.

EXPERIMENT	DIY MIND CONTROL
CONCEPT	Studying the effect of subliminal messaging on people by pelting them with visual cues toward specific behaviors and recording the results.
COST	$—$$
DIFFICULTY	⚙ ⚙—⚙ ⚙ ⚙
DURATION	☼ ☼—☼ ☼ ☼
DEMONSTRATION OR EXPERIMENT	This project is best done as an experiment.
TOOLS & MATERIALS	• Posters showing either specifically slow images or specifically fast ones • Timer • Pad/paper or a computer • Video camera (optional)

The term *subliminal advertising* was coined in the late 1950s by market researcher James Vicary to describe a process of flashing messages or images so quickly on a screen that they cannot consciously be picked up by the viewer. Vicary claimed that the subconscious recognition of the messages modified peoples' behavior.

While the results were later proven to be false, the idea of innocuous suggestions affecting actions is actually quite real. Pictures can easily alter mood, and mood will often color actions. The purpose of this experiment is to see whether an experimenter can use such visual programming to bend peoples' actions to the experimenter's will (or at least show some kind of recordable change in behavior).

Perhaps the most difficult challenge in this experiment is finding the right place to perform it. You need a public corridor where people walk from one identifiable end or point to another. You need to be able to watch people walking down the corridor in such a way so that you can time their journeys without it being obvious that you're doing so. Alternately, if there's one available, you could watch the corridor from a security camera, or set up a video camera to record people walking through the space. You also need to be able to get permission and access to hang posters in the corridor. You could conceivably use a school corridor, a store or mall, or other social meeting space.

What's My Hypothesis?

With this experiment, you really have two choices of hypothesis: Will the difference in images make people change how quickly they walk down the hall, or not?

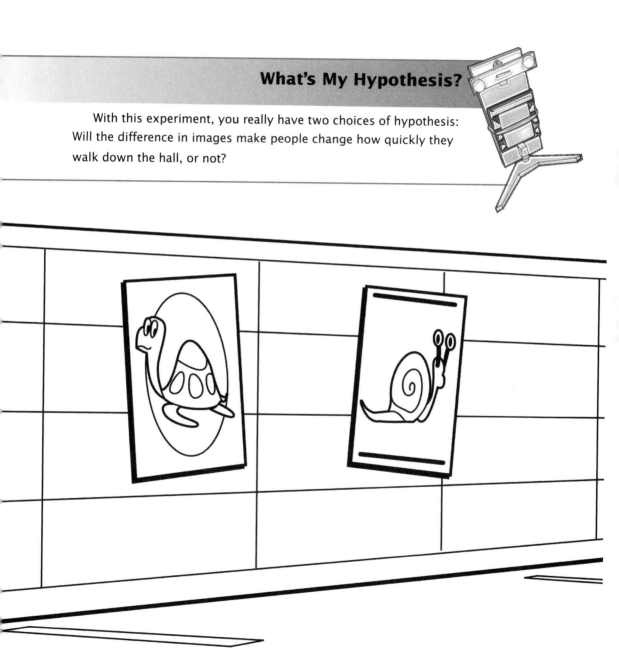

THE EXPERIMENT

STEP 1: Print out or acquire images of two types. The first set should be "slow" things: turtles, a glacier, and so on—whatever images you can find that easily represent the idea of "slow."

The second set should be "fast" things: rabbits, race cars, jets, sprinters—you get the idea. Basically, you want to be looking for images that convey the opposite of the first set of images.

STEP 2: On a given day, set up the "slow" posters along the corridor. Not too many, but not too few. They do not need to be the only posters or pictures in the corridor—you don't want it to be too obvious or call attention to them. Also pick two spots on either end (approximately) of the hall as the "start" and "end" points. If it helps, put a piece of tape on the floor or wall to mark the points so you'll be able to see them when you observe the corridor (but again, don't let them be too obvious or easily damaged/removed).

STEP 3: For a specific period of time, say for one hour between 1:00 P.M. and 2:00 P.M., observe the hall. At random, select people walking down the hall and time how long it takes them to travel from one tape mark to the other. Also record a brief description of the person and note whether they were doing anything else at the time they made the journey (talking to another person, talking/reading on a phone, eating food, etc.). Get as large and varied a sample set as possible (lots of people, and as many different as you can—old, young, male, female, different ethnicities, and so on).

STEP 4: You'll repeat the study again, but next time do it with the "fast" images on the walls. If possible, record your data in the same time period on the same day of the week, and avoid any significant changes like end of a school year, or holiday break. Make exactly the same kind of recordings for the same variety of people.

ANALYSIS

The number crunching here is pretty obvious: Compare the average speeds of the first set of subjects to that of the second set. Was the second group noticeably faster walking down the hall than the first?

You can go deeper, though. You recorded approximate ages, genders, and even ethnicities. Break out the results according to these categories and see whether specific subgroups showed greater or lesser affects from the stimuli.

If you have a real mathematical knack, you may also want to apply some statistical analysis to your raw data to try to isolate and remove outliers. It's perfectly possible that one person, running through the hallway completely oblivious to her surroundings, could impact the average to the point of showing a false trend. Statistics will help you overcome that kind of effect. Or you could make sure to only record data on people moving fairly normally. Just be sure you explain that in your report!

And so, my good aspiring mad scientists, you have now taken your first great step in understanding how to quell the masses who might otherwise rise against you when it comes time to make your "true" plans known. You can check one more thing off that "path to world domination" checklist we both know you're keeping, and raise a cold glass of milk in celebration.

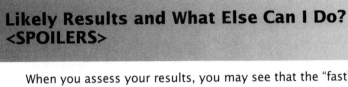

Likely Results and What Else Can I Do?
<SPOILERS>

When you assess your results, you may see that the "fast" images make people go faster, and the "slow" images slower. With these two results, you're actually pointing to a third data set: how people traverse the hall without specific stimuli being presented. For an even more robust experiment, you could actually record a third such set of data (the control) to compare to the other two.

You can try this experiment with other stimuli. For example, try putting "happy" posters on the walls and asking everyone who walked through how they felt, and then repeating with "sad" posters. Or you could try "hot" and "cold" images and then ask people to assess their temperature as they walked through the corridor you set up. Really, the sky (or the mind) is the limit.

</SPOILERS>

Growing Crystals for Power

Idea by Kathy Ceceri

Someday, we'll break free of our secret alien oppressors. Oh sure, nobody knows about them. They all look the same as we do, unless you catch them at the right moment when their eyes glow, and all of a sudden you hear background music playing menacingly—just like the movies.

Aliens are notorious for having cool devices and powers. One neat thing you see they often have in the movies or books is a *stargate*—a large device shaped like a ring, which when powered correctly, opens a portal through space to other such devices located on other worlds; other galaxies even. It's like a space door!

But to get a stargate to work, you need a controller mechanism to filter incoming power through special crystals. This experiment will show you how to grow your own crystals. It will power you through space and beyond!

EXPERIMENT	GROWING CRYSTALS FOR POWER
CONCEPT	Under the right conditions, atoms or molecules in a solution will start to arrange themselves in regular patterns, forming crystals.
COST	$
DIFFICULTY	⚙
DURATION	☼ ☼ ☼ ☼
DEMONSTRATION OR EXPERIMENT	Mostly a demonstration, but for an experiment, you can try different string materials, concentrations, or temperatures.
TOOLS & MATERIALS	• Alum (found with the spices in the supermarket; about $5 for 2 oz.) • ½ cup hot water • 2 clear disposable cups or jars (glass or plastic will work, as long as they are microwave safe) • Coffee filter or paper towel • Paper plate • Nylon fishing line or thread • Craft stick • Food coloring (optional)

We're mostly familiar with crystals that have the good press agents. Take snow—it has all those meteorologists out there doing daily updates on it all winter. Diamonds and other special gemstones are crystals as well, and are highly sought after. But there are many kinds of crystals, some of which we can even make ourselves.

Crystals are formed when atoms or molecules are arranged in a repeating pattern. Sometimes it's a matter of some combination of heat, pressure, time, or all three, but under the right conditions, atoms or molecules in a liquid will start to come together around a crystal nucleus in regular household mad scientist laboratory conditions.

A Little Bit of Science History

Crystals have been used in electronics for quite a while (almost as long as electronics as a thing has existed). In 1880, Pierre and Jacques Curie (they were brothers; Pierre was married to Marie, with whom he won a Nobel Prize for Physics; their daughter Irene later won a Nobel Prize for Chemistry with her husband, Frederic Joliot—smart family!) discovered that some crystals produce an electrical charge when compressed, which is known as the piezoelectric effect. Today, crystals are used to power watches, produce laser beams, as sensors, and to convert electrical energy into vibrations in radios.

To grow a mass of crystals on another surface, like a geode forms, you can pour a specific chemical solution over a rough surface, which gives the crystals places to grow. To grow one giant crystal, you can take a tiny seed crystal and encourage the atoms or molecules to grow around it. As the crystal gets bigger, it will always retain the same general shape—the shape in which the molecules themselves are arranged. So a crystal is just a magnification of what's going on at the molecular level.

And that's exactly what we're going to do. We will use alum, a substance made famous in old Warner Bros. cartoons for making you pucker and shrink when you accidentally put your head in a jar of it. Actually there are a few kinds of alum compounds, literally called alums. We want the most common form, potassium alum—hydrated aluminum potassium sulfate, $KAl(SO_4)_2 \times 12H_2O$—which can be used to grow large crystals. It is also used to dry and shrink vegetables like pickles, in styptic pencils to dry up shaving cuts, and in some recipes for homemade Play-Doh.

This experiment will show you how to grow a basic, white alum crystal. You can build a science fair–type of experiment out of this by making hypotheses based on any of the factors (time, heat, concentration, chemicals used) included in these instructions.

It is possible to make crystals of different colors using additives (including food coloring). You can even make crystals that glow under black-light. Researching possible additives and trying to produce crystals of different colors could be an excellent science fair project.

GROWING YOUR CRYSTALS

STEP 1: The first thing we need to do is make some starter crystals— the basic seeds of what our larger crystals will grow from. To make the starter crystals, we'll create a saturated solution of alum in water, from which the seed crystals can precipitate.

Pour $\frac{1}{2}$ cup of hot tap water into a clean jar. Add a little alum at a time, stirring constantly. Keep adding more alum until it stops dissolving and you start to see powder collect in the bottom of the cup or jar you're using. Place a paper towel or coffee filter on top of the cup and let it sit for a day or two in a place where it will not be moved or jostled. Yes, this is not a fast project.

STEP 2: When you see small crystals developing in the bottom of the cup, evaluate them for size. You need at least one, if not more, seed crystals that you can actually tie a piece of fishing line or thread around. When you think you have that, carefully pour the liquid in the cup into a second cup, leaving the crystals behind. Pour the tiny crystals out onto the paper plate.

Pick the largest crystals out to be your seed crystals. For each seed crystal, tie a string around it, leaving about 6 inches of string at the top. Nylon fishing line is best because the line itself will not start accumulating crystals. If you find it too hard to handle, however, regular thread works fine.

Tie the other end of the line around a craft/Popsicle stick and rest the stick across the top of the second cup so that the seed crystal hangs in the middle of the cup, without touching anything.

From the Mad Scientist's Notebook

If none of your crystals are big enough to handle when you first pour them out, you'll need to let them sit another day or two. You can start over without using more alum by pouring the small crystals back into the cup and putting the cup into the microwave oven for a few seconds. Heating the water should make them melt back into a solution. Then leave them as before and wait.

STEP 3: The hardest work you have to do for this experiment is to watch and wait, because it's all about evaporation. As the water in the cup evaporates, the alum solution becomes more concentrated, which causes more alum to go into solid crystal form (precipitate). Because we have this seed crystal in the cup, the alum will mostly decide to latch onto it and form up into the same nice crystalline structure it's already used to. However, if crystals start to form on the inside of the cup, you'll want to carefully pour the solution into a clean cup and move the seed crystal, or the solution will precipitate on the other crystals as well.

The crystal will keep growing until all the alum in the solution is used up. To keep the crystal growing, make a new solution of alum as in Step 1, being sure to let it cool to room temperature before moving the crystal. As the crystal grows, it will encompass the string supporting the seed crystal as well. Eventually the string will be coming out of the middle of the larger crystal.

In the end, your level of patience (and the size of the cup) will determine when the crystal is "done." You can keep growing it in this manner relatively indefinitely.

When you're finished growing your crystal, you can take it out of the water and snip off the string (if you want). The basic alum crystal will look like a short, fat diamond with the top and bottom points sliced off. The cool thing is that this is also the shape of each alum molecule.

And that's it. You have grown your own crystals and learned the skill to make more. Even better, you're on your way to powering up that old Space Portal and taking the battle with the aliens onto their own home turf. There's only room for one planetary ruler here on Earth, and that position will be filled by a mad scientist, gol-durnit!

Biosphere Breakdown

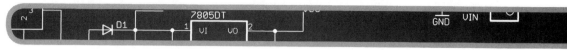

Idea by Tristan Russell

I n a moon base or space station, where life exists in a bubble that keeps air and pressure in and the ravages of solar radiation and the cold out, the biosphere is all. Or a biosphere can also be a dome, or a great big cave carved out of solid rock, where inside there is abundant life, carefully managed for temperature, humidity, oxygen/ nitrogen balance, light, and so on. The point is, a biosphere is a closed system, where everything inside lives in balance with one another. Plants, and maybe even animals (in the form of microbes in the soil or insects in the air to keep the plants healthy) live in a programmed symphony of existence.

Until something foreign is introduced. Something toxic.

And then, isn't it fun to see what havoc ensues?

EXPERIMENT	BIOSPHERE BREAKDOWN
CONCEPT	Study how toxins can affect an ecosystem by building your own biospheres and introducing chemicals into them.
COST	$—$$
DIFFICULTY	⚙ ⚙
DURATION	☼ ☼ ☼ ☼
DEMONSTRATION OR EXPERIMENT	Great for an experiment, using different household chemicals.
TOOLS & MATERIALS	• 4 (or more) 2-liter soda bottles • Scissors or shears that will cut plastic • Packing or duct tape • Bean seeds • Miracle-Gro planting soil • Water • Toxins (bleach, antifreeze, or Formula 409 or other household cleaner)

People have been dealing with the effects of chemicals on local ecologies for as long as there have been, well, people living with ecology. The issues became more pronounced during the Industrial Revolution when large factories started discharging the often-toxic byproducts of their work into local waterways, harming plants and animals alike. In the United States, regulations have done much to stem the most toxic effects of living near large manufacturing companies, but ignorance and accidents still cause tragedy from time to time.

It is as important to stop pollution as it is to understand its effects, so that we can learn to counter it. This experiment helps us measure what happens when we introduce a foreign substance into a biosphere of some kind. It is a really interesting ecological project and one that is perfect for a school science fair (it has received an A!).

Before you start, try to do some research on which chemicals are found in your local ground water. Look up the manufacturer or

From the Mad Scientist's Notebook

Okay, up until this point, I've been using the terms *biosphere* and *ecosystem* pretty interchangeably. And for the small scale we're dealing with, it works. However, we should be clear about the difference. On the planet Earth, for example, there are many ecosystems: deserts, marshlands, a plethora of forests, and more, each one distinct from the next. Indeed, the coniferous forest on one hilltop in one state is a completely different ecosystem from the one on the hilltop the next state over. In fact, the boundaries of an ecosystem are usually not very well defined, but in general encompass all the living things and the nonliving physical components with which they interact, in a given area. Simple, right?

On the other hand, a biosphere has very definable limits. For example, the earth is pretty much a biosphere. It is the sum total of all the ecosystems inside of its arguably closed boundaries. Yes, we do go into space, and meteors do add material to the planet, but for the purposes of most scientific study, nothing gets in and nothing gets out. The sum total of all matter inside doesn't change, and all biological processes occur without outside influence. Think of it like a bubble of biology, which is probably why it's called a biosphere.

And so, what we are creating for this experiment are a series of biospheres, each one encompassing a single ecosystem.

other industries (farms, golf courses) near you, and try to discover what kind of chemicals they might be using. You can often go to your state or county website, or see if there are local ecological watchdog groups keeping an eye on things.

Then, figure out which household products commonly contain these chemicals. Some substances that may be common in your neighborhood or home are bleach (used for your own laundry, or at clothing cleaners and Laundromats), antifreeze (from car repair shops or your garage), or Formula 409 and other household cleaners. You can purchase these items at the store or use what you have on hand.

As with other projects in this book, use care when dealing with chemicals. Wear gloves, safety glasses, and even a breathing mask, and work in a well-ventilated space. If you ever feel like fumes are causing you to cough or feel funny, leave the room immediately.

BUILDING THE BIOSPHERES

STEP 1: We're going create each ecosystem in a separate bottle (which makes it a biosphere, riiiiight?), so start by cleaning the bottles as well as possible. We don't want any preexisting contaminants in them.

STEP 2: To get the soil and other components of the biosphere inside the bottles, we have to cut them open. The most straightforward thing to do is to cut the bottles completely in half, crosswise, but that will make it harder later to close them back up. If you can, try only cutting them halfway and bending them open to fill them, then closing them back up afterward. That technique should be easier.

STEP 3: Poke a few holes in what will be the bottoms of your biospheres so they will drain when you water them. Fill the bottoms of the bottles with soil.

STEP 4: Plant your seeds in each bottle. Try to be consistent in the number and pattern of seeds in each one, and add identical amounts of water to each one. Follow the watering information that came with the seeds.

STEP 5: Label each biosphere, giving it a unique number and noting which toxin you'll put in each one. One must be left toxin free, to act as the control. Label this one . . . "Control!" Make notes on the conditions of the soil. Take pictures.

What's My Hypothesis?

To make this a science fair project, you need to have formed a hypothesis at the start, and just saying "all the poisoned plants will die" won't cut it. It's logical to assume there may be some effect, but how much? Will a particular toxin completely kill a given biosphere, or not? If you truly research the ingredients of a given chemical, you may be able to make a more educated hypothesis, and have more interesting results to analyze later. And you may end up surprised by just which chemicals are, and are not, bad for a biosphere.

STEP 6: Close up your biospheres, bring the halves together in whatever fashion works best for the way you opened them, and tape them closed (though don't use a lot of tape, because you're going to need to open them to water). Put the bottles in the sun and let the beans sprout (if you're lucky, other things might grow, too).

During the first week, observe the plants every day, water every other day, and make sure the plants are in a sunny but not too hot place.

STEP 7: During the second week, add 1 cc of toxin to each of the experimental plants once every two to three days, delivering through a dropper onto the soil. You can pick the same spot each time, or a different spot, but be consistent in each biosphere. Observe, and at the end of week two, take pictures.

STEP 8: In week three, just repeat week two.

STEP 9: And in week four, repeat week two one more time. At the end of the week, look at which plants are still growing and green, and which look different. Also, some of your ecosystems might be

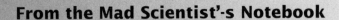

From the Mad Scientist'·s Notebook

Okay, I called these biospheres, and I'll be honest, they're really not. I know, I know, we're working at scale, and nothing's a perfect representation, but one other very salient point kills the whole biosphere analogy here: We're watering the sprouts. A true biosphere is a closed system and would have its own water cycle. We're not quite that good at biosphere-building, so we are taking a few shortcuts. But the science is still sound!

more hospitable to outside life than others, so take note if any insects have taken up residence, or if you brought in some other plants that you weren't aware of at the start (perhaps there were unsprouted seeds in the soil you used).

You should take notes regularly in addition to the pictures, because you can note smells and other effects that are not easily observable. If your beans do not sprout, you can try pre-sprouted plants instead. Any plants will do.

Also, don't mess with the ecosystems too much. Mold, insects, and other organisms may decide to live in your ecosystem. That's natural. Don't add excessive water. Water will condense on the side of your plastic bottles, creating a greenhouse effect within the bottle.

ANALYSIS AND CONCLUSIONS

So, what happened? At the end of your experiment, that's what you need to be able to answer. Organize your observations and images so you can see the progress of each plant over time, especially in connection to the introduction of toxins to each biosphere. Answer questions like "did the plants start dying right away, or did it take a specific total amount of toxin to begin having an effect?" Pay attention to anywhere the plants weren't affected at all. Why might that be?

Through this, you'll have a better understanding of how the chemicals we use every day can affect the life all around (and including) us; maybe not much, or maybe quite seriously. And knowing this, you'll be in a better position to act in the future. Of course, *how* you act is entirely up to you.

Can You Dodge a Laser?

Idea by Dave Giancaspro

I n space, many things can rely on reaction time: hitting the button to close the airlock after the alien is sucked out but before you are sucked out with him; timing the warp engines just right so you can slingshot around the sun and travel back in time; using your force powers to block blaster fire from an annoying smuggler. All of these things require sharp reflexes and a steady hand. Which is why, as an aspiring space-bound mad scientist, making sure your reflexes are up to snuff, and even perhaps finding ways to improve them, are all-important parts of the job.

And the first step toward testing reflexes is building the tool to do it!

EXPERIMENT	CAN YOU DODGE A LASER?
CONCEPT	Build your own reaction-time tool with a laser pointer and some home-hacked electronics.
COST	$$—$$$
DIFFICULTY	⚙ ⚙ ⚙—⚙ ⚙ ⚙ ⚙
DURATION	☼ ☼ ☼
DEMONSTRATION OR EXPERIMENT	Building the tool is an excellent demonstration of electronics assembly and programming with the Arduino open-source platform. The tool can then be used in a variety of experiments.
TOOLS & MATERIALS	• Laser pointer • Arduino • Computer (assumed to be already owned) • 2 1K resistors RadioShack Part # 271-004 • 10K resistor RadioShack Part # 271-006 • Photoresistor—RadioShack Part # 276-1657 • Green LED RadioShack Part # 276-304 • Red LED RadioShack Part # 276-209 • SPST pushbutton switch RadioShack Part # 275-1547 • MPS222 transistor RadioShack Part # 276-2009 • Project enclosure RadioShack Part #270-1801 (optional) • LED holders—RadioShack Part # 276-079 • AAA battery holder RadioShack Part # 270-398B • 2 AAA batteries • Ballpoint pens with opaque barrels • Duct tape (of course) • Electrical tape • #8 nut and bolts • 16 AWG stranded wire (this can be salvaged from an old CAT 5 networking cable) • Soldering iron • Small saw or Dremel tool

This experiment's title asks whether you can dodge a laser. The very simple answer to that is no. There's no question about it, what-soever. A laser is light. Light travels at the speed of light (duh), which is very fast. You just cannot move fast enough to dodge a laser—

human reaction times are measured in units so much larger than the time required. But we can still use a laser to test our reflexes, because while we can't actually get out of a laser's way, we can measure how long it takes us to move once we see the laser.

For this experiment, we're going to do some pretty neat electronics hacking. We are going to build a reaction timer that uses a laser, a light sensor, and an Arduino board with some special programming to measure how quickly a person can react to visual stimuli.

Please note: This is a pretty advanced experiment, requiring soldering and some work with programming. Don't be afraid of it! This is some really cool science and engineering, and we all learn by doing. That being said, gauge your own and your kid's aptitude and experience with these scopes and honestly evaluate whether you're up to it. If so, great! This is the kind of science fair experiment that can knock people's socks off.

What Is This "Arduino" of Which You Speak?

I introduced readers to the Arduino open-source programmable circuit boards in my first book and used them again in the second. Arduino is a very cool open-source initiative that puts practical programmable electronics into the hands of hobbyists. Check out www.arduino.cc for some great information about Arduino.

The boards can be built by hand or purchased preassembled, and the software can be downloaded for free. You can buy an Arduino board, hook it up to your home computer, download and run some free software, and actually program the chips on the board to do different things with modules you attach to the board. An Arduino board will help you teach your child that all the chips and wires crammed into every piece of home electronics you own aren't really magic boxes but instead are simple devices that you can easily learn how to hack with the right tools.

The components of this reaction timer include:

Laser: We are going to hack a common laser pointer so we can turn it on and off with a switch that is connected to the Arduino board and our computer. This will help us be able to count the time between when the laser turns on and when the sensor sees it.

Sensor: We'll just use a photoreactive sensor (an electronic component) that we'll wire up to our Arduino board. We'll calibrate the system by checking that the sensor can see the laser. Then, when we go into testing mode, the test subject will hold his hand between the unpowered laser and the sensor. The test supervisor will trigger the laser (and the timer), and the subject will move his hand out of the way as quickly as possible when he sees the light hit it. When the sensor sees the laser, it will turn off the timer.

Controls: The controls for this tool are simple: a red light-emitting diode (LED), which indicates when the sensor cannot see the laser; a green LED, which lights when the sensor can see the laser; and the switch for the test supervisor to use to trigger the test. If you want to "pretty-up" the whole build, you can mount these into the optional project enclosure.

Arduino Board: This is the brain of the operation. All the components will be connected to this board, which will in turn be connected to your computer.

Computer: Your computer will load the testing code onto the Arduino board, and then act as the visual output for the test results.

WIRING THE SENSOR

STEP 1: Take one leg of the sensor and solder it to one side of the 10K resistor.

STEP 2: Solder a 3-foot-long piece of wire to the junction of the resistor and the sensor. Insulate the connection with electrical tape. Connect the other end of the wire to the Arduino "Analog 0" port.

STEP 3: Solder a 3-foot-long piece of wire to the other sensor leg and insulate the connection with electrical tape. Connect the other end of the wire to the Arduino "5V" port.

STEP 4: Solder a 3-foot-long piece of wire to the other leg of the 10K resistor. Connect the other end of this wire to the Arduino "GND" Port.

STEP 5: Take one of the pens apart and cut the barrel in half.

STEP 6: Slip the leads for the sensor into one end of the barrel and push them through so they come out the other end. Slip the sensor into the end of the barrel so it fits snugly and points outward. Apply tape as needed to keep everything secure.

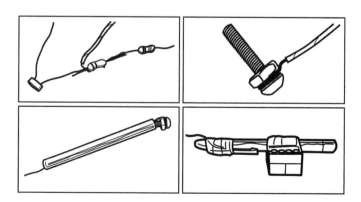

HACKING THE LASER

The laser needs to be hacked so we can control it with the Arduino board. In order not to destroy the laser pointer, we are going to make an adapter so the batteries can be on the outside of the pointer and we can control the on/off with our own switch.

STEP 1: We need to start with the laser pointer on. If it can be switched on, do that. If it uses a push-button mechanism that turns it back off when you release pressure, you need to use the duct tape, and perhaps something hard taped over the button to keep the button in the on position.

STEP 2: With the laser pointer still on, open up the back and remove the batteries.

STEP 3: Take a 2-foot length of wire and strip off around 2 inches of insulation from one end.

STEP 4: Wrap the bare wire around the #8 bolt and screw on the nut to make it tight to the head.

STEP 5: Disassemble the other pen barrel.

STEP 6: Place the empty pen barrel into the laser pointer's battery compartment and cut the barrel off so that about $\frac{1}{4}$ of an inch is left sticking out of the laser pointer. Take this piece back out.

STEP 7: Slide the wire connected to the bolt up through the barrel so it's sticking out of the other end.

STEP 8: Slide the nut/bolt/wire combo into the pen barrel so the wire comes out the bottom, and the bolt head "caps" the top.

STEP 9: Tape the nut in place so the nut is completely covered but the head of the bolt is exposed. You'll end up with a pen barrel that has a wire coming out of one end and a bolt secured at the other end. This is our replacement for the batteries that we removed and will allow us to run current into the laser pointer from our external batteries (which we can control with our Arduino board).

STEP 10: The other half of the circuit that will get power to the laser pointer is pretty easy to make. You just need to find a bare metal part on the outside of the pointer. The best way is to scrape the paint off the outside of the pointer casing.

STEP 11: Get another 3-foot length of wire and remove around 2 inches of insulation on one end.

STEP 12: Wrap the bare end of the wire around the pointer case where you exposed the metal. Be sure to make good contact with the bare metal.

STEP 13: Tape the wire in place, and then strip off 1 inch of insulation from the other end of the wire.

STEP 14: Put the pen barrel into the battery compartment of the laser pointer nut-end first so it makes contact with the spring at the bottom. Then strip off 1 inch of insulation from the other end of its wire.

STEP 15: Put two AAA batteries in the battery holder.

STEP 16: And now we'll test that we got everything right. Hold the red (positive) lead of the battery pack to the wire coming from the laser pointer exterior case, and the black (negative) lead to the wire coming from the pen barrel. If everything is right, the laser pointer should light, indicating that we successfully externalized the power source. If not, check the wiring, and make sure your button-lock is still working.

STEP 17: Once you're sure everything works correctly, tape it all up to keep things secure (but do not tape up the battery connections—those will be hooked up to other parts).

WIRING THE CONTROLS

Optional Box

This stage of the building instructions is partially optional. If you'd like to, you can use an electronics enclosure box to mount and secure the lights and switch that make up the controls. You do not have to do this, and instead you can have the controls connected to each other, but loose. It's less tidy, but quicker.

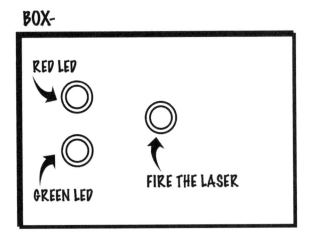

STEP 1: (Optional if you use the electronics enclosure box) Take the red and green LEDs and mount them in the box using the LED holders. You may need to drill holes into the box for each of the

components; reference the parts to determine the hole diameters required

STEP 2: Solder a wire to the long lead of the green LED and connect this wire to the Arduino port "D3."

STEP 3: Solder a wire to the long lead of the red LED and connect the Arduino port "D4."

STEP 4: Mount the switch in the box (also optional).

STEP 5: Solder a 1K resistor to one side of the switch, and solder a wire to the other side of the resistor. Connect this wire to the Arduino "5 Volt" connection.

STEP 6: Solder a wire to the same switch terminal that's connected to the resistor, and solder the other end of the wire to the Arduino port "D7."

STEP 7: Solder the other switch terminal to the short leg of the red LED.

STEP 8: Solder the two short legs of the LEDs together and solder another wire to this connection.

STEP 9: Run the wire you just soldered to the Arduino "GND" connection.

WIRING THE TRANSISTOR

Transistors are like little switches. When you apply current to one terminal, a larger current flows through the other terminals. This will play a key role in our tool.

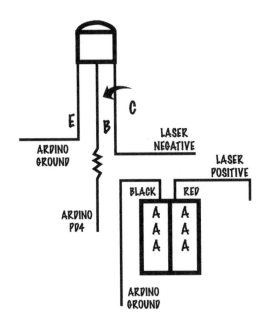

STEP 1: Solder one end of a 1K resistor to the middle lead of the transistor (also known as the base).

STEP 2: Connect the other end of the resistor to the "D5" port of the Arduino board.

STEP 3: Connect the positive side of the laser pointer (the lead from the exterior case) to the red (positive) side of the AAA battery pack.

STEP 4: Solder the negative side of the laser (the lead from the bolt) to the transistor terminal that's on the right when the flat side of the transistor is facing you (aka the "Collector").

STEP 5: Connect the "Emitter" (the left-hand terminal of the transistor when the flat side is facing you) to the "GND" on the Arduino board.

STEP 6: Connect the negative (black) lead of the battery pack to the "GND" of the Arduino board.

MOUNTING THE LASER AND THE SENSOR

After all this electronics work, we actually get to one of the most difficult technical challenges of building this tool: making sure the laser and the sensor can align with each other. After much trial and error, I developed a very simple method: Mount each of them on bases made of LEGO bricks to hold them a couple of inches off your work surface, and a couple of feet away from each other horizontally, and then tape them down so they won't move. The point is that the laser has to shoot straight into the sensor, and you need to be able to rest your hand in between to block the laser from hitting the sensor.

FIRING UP THE CODE

The code we'll use to program the Arduino board has three simple functions:

1. Align the laser—the Arduino checks the sensor to make sure it can see the laser when it's turned on.

2. Fire the laser—fires the laser after you push the button.

3. Display the time—displays the time between the laser firing and the sensor registering it (meaning pretty much how long it takes to get a hand out of the way—the reaction time).

To display the data, we use the Serial Monitor of the Arduino Sketch Software (this is freeware, available at www.arduino.cc). When you start to run the code, go to Tools → Serial Monitor, which will display the Serial Monitor. Make sure the Arduino programming cable is connected (it uses a USB cable between the Arduino board and your computer) to see the messages.

STAGES OF REACTION TESTING

1. If the laser is aligned, you will see a message and the green LED will be on steadily.

2. Place your hand in between the laser and the sensor. The serial monitor will say "hit the fire button."

3. Once the test supervisor hits the fire button, the red LED will light and the laser will fire. The subject must move his or her hand as quickly as he or she can.

4. The serial monitor will display the reaction time.

EXPERIMENT IDEAS

It takes some work to build this new tool, but that's okay, because you can use it for a wide variety of potential experiments. Some of the questions you can ask and answer through experimentation include:

1. Does caffeine affect reaction time? You can test your subject before and after drinking coffee, soda, or other caffeinated drinks.

2. Does playing video games affect reaction time? Test subjects' reaction times, then have them play up to two hours of video games. Test them again to see if the conditioning from game playing changes their times.

3. Is your left hand faster than your right? Test a number of subjects on either hand. Make sure to note who is naturally right- or left-handed.

4. Does the testing itself affect reaction time? Test subjects daily over a 1-week period. Do their times get better? Is there an upper limit to their reaction times?

Note

Always remember to include a "Control" in your tests—one person who doesn't do the potentially affecting behavior so you can rule out other environmental conditions from your results. This is vital scientific methodology!

Your imagination is pretty much the limit with this tool. It's great for your arsenal of mad scientist equipment. Whether you're training your elite squad of cyborg ninja astronauts, or improving your own ability to catch meteorites with your bare hands, you can't go wrong with a laser reaction-timer. Heck, you can't go wrong with a laser-anything!

Spaceship Design
Building Your Own Wind Tunnel

Idea by Dave Banks

Space: the final frontier. We've come a long way in recent decades and one of these days we may find ourselves flitting between solar systems like we take a drive downstate to Grandma's house. But technology often requires baby steps (as much as mad scientists try to take short cuts sometimes). And though we have built devices that can launch us off the planet, and even land us on the moon, if we want to go to other planets, our spaceships (or death gliders, or spy drone aircraft) all need to be designed for one very important thing: the ability to fly within an atmosphere. They need to be aerodynamically sound. And to study aerodynamics—the physics of how things work when zooming through the air really fast—it's always really handy to have one thing: a wind tunnel.

EXPERIMENT	SPACESHIP DESIGN: BUILDING YOUR OWN WIND TUNNEL
CONCEPT	Build your own wind tunnel at home, and use it to study aerodynamics.
COST	$—$ $ $ $
DIFFICULTY	✿ ✿ — ✿ ✿ ✿
DURATION	☼ ☼ ☼ — ☼ ☼ ☼ ☼
DEMONSTRATION OR EXPERIMENT	More useful as a proof of concept and use for demonstration. However, once built, it could be used for a variety of experiments.
TOOLS & MATERIALS	• Several large pieces of cardboard • 7" × 24" galvanized sheet metal furnace pipe • Duct tape • Metal shears • Safety goggles and gloves • Tinsel, newspaper, or other thin paper • Wire hanger • Table or floor stand fan • Egg crate or similar plastic grid screen • Sheet of clear acetate or heavy, clear plastic toy plane/foam wing, or other object to be tested.

Ever since Sir Isaac Newton first posited a theory of air resistance in book 2 of his *Philosophiae Naturalis Principia Mathematica* (indeed, Newton himself supposedly tricked out his buggy with spoilers for racing through the countryside . . . not), man has been fascinated with making things sleeker and more streamlined. Aerodynamics is largely concerned about a single force: drag. Drag is defined as the force that resists the motion of an object through a fluid. And yes, air is a fluid when we're talking about aerodynamics; and most of this science applies to bodies traveling through water, too. But it's harder and messier to build a water-tunnel, so we'll save that for another book.

You can think of drag like a kind of friction. If you try to do a moonwalk on a rough concrete surface, you pretty much won't

From the Mad Scientist's Notebook

What is a wind tunnel? Well, uh . . . it's a tunnel. With wind. Can we move on?

More than that, a wind tunnel is one of the major tools an aerodynamic engineer uses to study air-flow around an object. A wind tunnel is typically made up of just a few components (though the parts of a professional wind tunnel are far more complex and powerful): a fan to power the wind, a settling chamber or entrance cone to straighten airflow, a test section where the object is observed, and a diffuser to slow the exiting air. In our case, we aren't dealing with high velocity air, so we are skipping the diffuser.

And we'll be using a couple of words a lot in this experiment: turbulent and laminar. Most of you will have heard turbulent before and understand the basic idea from flying on airplanes. When there is turbulence in the air, planes bounce around. What that really means is that the plane is flying through turbulent air—air that is blowing around in a messy, chaotic manner such that it shakes the plane. Simple, right?

So laminar, a less familiar word, is just the opposite of turbulent. You can get a really good idea of what laminar versus turbulent is in your kitchen. Fill a 2-liter bottle with water. Turn the bottle upside down over your sink and let water come out. You hear how much noise it's making? How disturbed it seems? How . . . turbulent? Okay, stop. Now gently angle it, like you're going to pour a glass, and just let a trickle come out. It's quiet and smooth, right? That's laminar flow. And now you know!

move. But if you practice your *Risky Business* slide on a freshly polished wood floor, you'd best be wearing elbow pads and a helmet, because the friction will be so low between your socked feet and the ground.

By creating shapes that minimize drag (a kind of friction in the air), engineers can create lift—the upward force on the underside of a wing, for example, that helps support airplanes in the sky. Or they can create the opposite of lift, called "down force," which pushes down on a race car's extremely light-weight body to keep it from flipping over at 200 mph.

Automobile companies, racing teams, airplane designers, and manufacturers of all kinds spend untold billions on aerodynamics tests each year. Building a proper wind tunnel for these tests costs anywhere between \$25–\$80 million to construct and can cost as much as \$10,000 an hour to run. As fledgling mad scientists, our budgets run a little lower. We're going to create our own wind tunnel and learn about the basics of aerodynamics for under \$25.

BUILDING OUR WIND TUNNEL

Before beginning this project, turn on your fan. Move a piece of ribbon around the face of the fan. See how the ribbon reacts differently at different spots on the fan? This is because the air is moving about turbulently—the air is behaving like the crowds at the running of the bulls in Pamplona, Spain: running around all willy-nilly, bumping into other air, and trying not to get gored by the bigger air with horns. So when you hold your ribbon, some air is pushing it in one direction, some in another, and so on. To build a good wind tunnel, we must do our best to "straighten out" the air. We'll start by building a settling chamber or entrance cone, which will channel the air in one direction and get it moving in an orderly manner. Then we'll build the actual testing chamber where you'll perform your experiments.

STEP 1: From cardboard, cut out four trapezoid-shaped pieces, measuring 7.25 inches at the top and 18 inches at the bottom, with sides about 25 inches long. Use duct tape to tape the four sides together, creating a framework that looks a little like a pyramid with its top cut off. You should have one open end that is a square measuring 18 inches on each side, and the other open end that is a square measuring 7¼ inches on each side. It's going to act sort of like a funnel. More turbulent air will go in the larger end and get squashed down and straightened out until it comes out the smaller end in a smoother (more laminar) state of flow.

STEP 2: At the entrance (wide end), cut a piece of egg crate (or a similar plastic grid) to size and affix it to the entrance with duct tape. The grid will help straighten out our air. As a variable in testing, you can place a second piece of egg crate 4 inches from the narrow end of the chamber (actually, if you want this, you'll have to install it before you do the piece at the entrance).

From the Mad Scientist's Notebook

Just to be clear, when we say "egg crate" we're not talking about an egg carton. We're talking about the plastic crates eggs (or milk) are sometimes shipped in that have sides with a plastic grid. The grid is usually anywhere from ½ inch to 1 inch square, and usually it is also deep—maybe ½ inch. You could also find similar materials at your local plastics store, or even your local hardware store. There are types of diffusers/covers for fluorescent lighting that match the idea as well. Basically, we want the plastic grid to act kind of like a strainer for the air being forced through it, helping it redirect into mostly the same vectors for movement through our entrance cone.

STEP 3: Next, we'll create the testing chamber, where all the experimentation will happen. Take the 7-inch × 24-inch piece of galvanized furnace pipe (aka stovepipe) and separate the pipe along the seam by pressing down slowly and gently on the seam until the pipe unrolls.

STEP 4: Carefully cut a hole in the pipe with the sheet metal shears, about 8 inches by 6 inches halfway down the easier edge of the seam to cut. Whoever is doing this (whether it's you or your younger aspiring mad scientist who is of sufficient age and strength to cut sheet metal) must be wearing safety gloves and glasses when doing this work. Sheet metal is *extremely* sharp, and this is a potentially dangerous task.

Rejoin the seam of the pipe by tucking the edges back together. When the pipe is rejoined, cover all exposed, sharp edges with 2–3 layers of duct tape.

STEP 5: Drill a $\frac{1}{4}$-inch hole in the furnace pipe opposite the viewing hole you just created. This will allow you to control objects from outside the pipe. Tape a piece of clear acetate over the viewing area so you can observe your testing without allowing air to escape out the side (and become more turbulent). You could alternately do something cool by holding the acetate in place with magnates, since the sheet metal is metal, after all.

STEP 6: Finally, you'll be ready to join the entrance and viewing chambers by inserting the furnace pipe into the narrow end of the entrance chamber. Well, almost. The challenge is that everything needs to be at the same horizontal elevation.

The fan needs to point directly into the wide end of the entrance chamber, which then feeds into the furnace pipe. Because the furnace pipe is so much narrower than the entrance chamber, it needs to be propped up on whatever your work surface is, with some blocks or boxes to support the pipe in line with the center line of the entrance chamber. You could build a framework for it by cutting holes out of a box and feeding it through. Or you could even suspend everything from the ceiling!

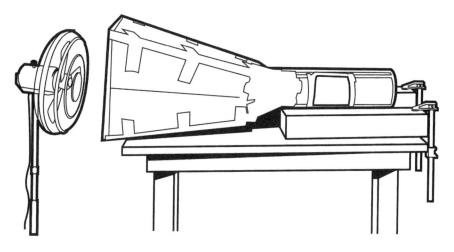

STEP 7: Tape the two pieces together using (you guessed it) duct tape. Try to eliminate gaps as much as possible to keep any air from leaking out.

READY FOR EXPERIMENTATION

Now you're ready to find a test object. You can use a toy plane, shape a wing from foam, or locate any similar object that can fit in the testing chamber. Tinsel is the best material to use for visualizing the flow of air, but you can use ribbon or small strips of newspaper or onion paper, too.

Tape a piece to the trailing edge of your test object, then figure out how to mount the object in the testing chamber. You'll want the object suspended in the center of the pipe so the air can pass freely around it. The way to do this is to use a wire coat hanger, cut to size and purpose, affixed to your test object and threaded through the hole on the other side of the pipe. This will also allow you to rotate the object while the air is on.

With everything in place, switch the fan on and observe the tinsel on your object. Does it trail horizontally? What happens if you rotate your object several degrees clockwise or counterclockwise? Try changing the speed of your fan. Try substituting your leaf blower (making sure all loose materials are well away from the other end of the tunnel). What happens if you remove the egg crate? Do you have a fog machine? Place it between the fan and the entrance chamber. Introducing fog/smoke can help visualize how air moves over an object.

So, now that you have an awesome testing device, you should figure out what to do with it. Sure, this makes a great demonstration project, but you can use it for proper experiments as well. You can study the effects of shapes on airflow by modeling different cross-section with foam or clay. Try building a series of car bodies using modeling clay, and then run them in the wind tunnel to see which one passes air over more smoothly. Refine your models and keep trying until you come up with an optimal shape.

You can also study surface drag with different materials, like paper, rough wood, polished wood, and more. Use similar shapes (maybe one car-body you perfected from the above paragraph), but make the primary surface of each one out of the testing material and compare how the air passes over each one. Will a paper surface be more aerodynamic that a wood one?

You can even model the effects of hurricanes. Build a scale-model house out of approriate materials and see how heavy winds affect the construction. If you can purchase an anemometer (the whirlygig spinning thing that measures wind speed), you can find out how fast the wind is going in your tunnel and then test your scale models at scaled speeds (if you have a 1/10 scale house, then a 10 mph wind in your tunnel would equal a 100 mph gale).

With your very own wind tunnel you can test the aerodynamics of almost anything you can think of. The sky's the limit. Or, rather, the wind.

INSIDE THE MAD SCIENTIST'S KITCHEN

Making Your Own Topsoil

Idea by Edmund Williams

The mad scientist is a study in contradictions. For example, the road to world domination often runs past many the salted field or rebellious city melted to glass via orbital death ray. Maintaining power often requires wanton destruction, and we have to be . . . okay with (if not enthusiastic about) that.

On the other hand, being a successful mad scientist means being the smarter, more efficient ruler. After all, isn't that the whole point to any reasonable bid for world domination? We know we can run the world better than everyone else. And so, a clear understanding and appreciation of recycling and the principals of reuse is key in keeping operating expenses down while maximizing available materials.

This is nowhere truer than in the kitchen, where every scrap of leftover food can and should be utilized (or reutilized) for its vital components. In the perfect mad-scientist-controlled society, nothing will go to waste (upon penalty of death! After which you will be recycled!). And to perfect our strategies for this brave and beneficial future, we much research the best ways to take care of certain types of household scraps.

EXPERIMENT	MAKING YOUR OWN TOPSOIL
CONCEPT	Use mushrooms and worms to generate a rich soil from old paper, coffee grounds, and other used plant materials.
COST	$ $
DIFFICULTY	⚙
DURATION	☼ ☼ ☼ ☼ (This is a long-term project.)
DEMONSTRATION OR EXPERIMENT	Demonstration or Experiment
TOOLS & MATERIALS	• 5-gallon plastic bucket • Trowel • 5- or 10-lb. oyster mushroom kit • Shredded newspaper or other scrap paper • Coffee grounds • Red wiggler worms • Seeds for new vegetables

At its heart, this project is about composting (actually, vermicomposting, which we'll explain later)—deconstructing waste vegetable matter via the biological processes used by fungi and worms, and turning it into a growing medium for new vegetable matter.

Part of most soils is broken down rock, but not all of it is. The rest is usually the product of the breakdown of vegetable and animal materials into their base organic compounds by microbes, fungi, and invertebrates. In this experiment, we are going to turn items that used to be plants into good quality topsoil, using scrap materials from your home. To do so, we will be using two separate organisms to complete the process.

The first organism is oyster mushrooms. Mushroom-producing fungi are excellent at breaking down woody, fibrous materials. They do so by growing white, cottony threads, called mycelium, into the material they are going to digest. This takes care of most of the decomposition and what is left is easily digestible for the second organism: earthworms. Each worm will eat half of its own body weight every day, turning what is left of the newspaper and coffee grounds into a rich dirt called vermicast (hence vermicomposting), or earthworm castings (yes, it's earthworm poop; get over it).

At the biological level, the organisms eat the oxygen, carbon, hydrogen, and nitrogen-rich materials in the scraps and turn them into different compounds that are especially useful in plant growth.

From the Mad Scientist's Notebook

What's going in the mix? Oyster mushrooms are actually quite flexible about their growing medium. You can add a number of things to the bucket as food for the mycelium. Just about anything that was once plant material can be used, especially sawdust, grass cuttings, old salad fixings, and so forth. It is particularly neat to add 100 percent cotton clothing that is worn out, just to watch it disappear. But steer clear of table scraps that may rot too quickly and cause contamination, or anything made of pine or other conifer wood. Dried-out egg shells are good, and fruits, vegetables, coffee grounds, and used tea leaves. Just no meat, dairy, or other fatty materials.

Also, as the volume of material in the bucket reduces significantly as the worms do their work, you can add small amounts of more kitchen scraps. It will take longer for the worms to finish, but you will get more dirt out of it.

STEP 1: Preparing Your Mushrooms

Buy an oyster mushroom (Pleurotus ostreatus) mushroom kit. These are commonly available online and usually come in the form of a 5-to-10-pound block of sawdust with mush-

room mycelium growing on it (see, for example, www.wheatgrasskits
.com or www.ufseeds.com), or inquire at your local nursery or organic
food store).

Follow the directions to get the kit to produce a flush of mushrooms. This step not only allows you to see the mushrooms growing
(a great experience for a child), but also gives you a good idea of
what the mushrooms are supposed to look like. If you would like,
you can wait a few weeks (again, follow the instructions on the kit)
and get a second flush out of the mushroom kit. Don't try for a third,
though, as the kit may have lost too much energy and may be too
weak for the experiment.

STEP 2: Setting the Mycelium to Work

Within a day or two of harvesting the mushrooms, you will need
to start your experiment. Take a 5-gallon plastic bucket (a 10-gallon
aquarium would be extra nice and allow great visibility for the experiment but would probably be more expensive unless you have
one lying around).

Fill the bottom of the bucket with a few inches of shredded
newspaper and a few cups of coffee grounds, making sure to add the
coffee filter as well. Break up the mushroom block with your hands
and sprinkle it on top of the newspaper and coffee grounds. Place
the lid lightly on top of the bucket and wait 2–3 days.

STEP 3: Slowly Adding Materials

After 2–3 days, the mushroom block should start looking white
and fuzzy. This is the mycelium (the body of the mushroom) growing out and looking for more food. At this point, add another handful of shredded newspaper and more coffee grounds.

The mycelium will start trying to colonize the new additions. To
keep just ahead of the growing mycelium, add the coffee grounds
from your daily pot every day and add a handful of shredded newspaper every other day. Keeping a very small amount of uncolonized

material in the bucket at one time will help keep contamination (usually mold) to a minimum. Stop when your container gets full.

STEP 4: Producing Mushrooms

Once the container is full, give the mycelium about 2–3 weeks to grow. Watch it to make sure it colonizes all of the newspaper and coffee grounds. About 2 weeks after everything is colonized by the mycelium, it is time to initiate production of mushrooms.

Move your experiment to a cooler location. A temperature of 60 degrees Fahrenheit is ideal, so a basement or garage might work well. Also begin misting the mushrooms with filtered water 3–4 times a day. Within a few days, the mushrooms should begin to develop primordia, or tiny mushrooms.

Refer to the directions that came with the initial mushroom kit for further information. Once you have harvested the mushrooms, leave the block alone for 3-4 weeks and initiate mushroom production again. You can do this 3 or maybe 4 times before the block is spent and won't grow any more mushrooms.

STEP 5: Adding Worms

Once the mushroom block is spent, it is time to add the earthworms. Red wigglers are the best for the job and again, amazingly, you can buy these online and have them shipped to you (try www.unclejimswormfarm.com).

Just dump them on top and let them find their own way down in. The more worms you buy, the quicker they will do their work of turning the remnants of the coffee grounds and newspaper into rich, black dirt. The whole process should take about 1–2 months.

STEP 6: Growing Vegetables in Your New Dirt

When the contents of the bucket have turned to black dirt, you can add seeds and move the bucket to a sunny location. What was once household waste can now be used to grow more food. Obviously this is a very long-term experiment: weeks to produce the mush-rooms, months to turn the waste into soil. But it's really, really good soil. On top of that, if you get yourself into a cyclical process with 3 or more buckets going in offset phases, you can create a nearly self-sustaining growing setup.

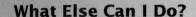

What Else Can I Do?

From this basic description, it's also easy to devise various science projects (keeping in mind the time it takes for any given part of this process):

- ▶ You can try different mushrooms, different worms, or different scrap materials (papers or coffee types).
- ▶ Alter temperatures.
- ▶ Play with the timing of certain transitions.
- ▶ Will a larger bucket help?
- ▶ Can you improve the process by adding other nutrients?

Ultimately, you want to maximize the growing capacity of your final soil product. To do so will require experimentation as well. Try testing each soil produced using different composting processes by growing the same vegetable in it and watching for variations in growth speed or volume of growth, making sure to include a control subject using some commonly available topsoil.

And with that, you can establish your agrarian policy for the new mad scientist world order. Remember, it's organic!

Growing Tasty Sea Monsters

Idea by Kathy Ceceri

magine a terrifying creature that has lived in the deserts for hundreds of millions of years. It will eat pretty much anything you put in front of it, vegetable or animal. It has three eyes, so it can literally see out of the top of its head. And it is one of the most resilient creatures ever, able to go dormant and survive without water for vast spans of time, only to awaken at the first touch of moisture, revive itself, start eating again, and start reproducing.

It's the stuff of nightmares.

But these creatures are also really cool and we can learn a lot from them. What's more, as a mad scientist, you can raise them. You can breed them. And you can train them to do your will. They can be your very own mindless eating-machine-sea-monsters/cuddly family pets.

And, in a pinch, you can eat them in a nice garlic and butter sauce.

PROJECT	GROWING TASTY SEA MONSTERS
CONCEPT	Observe the entire life cycle of an ancient shrimp species in a month or less.
COST	$
DIFFICULTY	☼ (Must be responsible enough to care for a pet)
DURATION	☼ ☼ ☼ ☼
DEMONSTRATION OR EXPERIMENT	Demonstration or experiment. Can test how conditions affect triops' growth, or whether triops can be trained
TOOLS & MATERIALS	• Triops kit with eggs, nutrients, and food pellets (available at museum gift shops, toy stores, or www.triops.com) • 1–2 gallons non-chlorinated water • Small- and medium-size clear plastic containers, such as a take-out or salad bar containers (rinse with bottled water) • Clean sand (rinse with bottled water) • Aluminum foil • Small thermometer • Gooseneck desk lamp with incandescent bulb (for warming, if needed) • Ingredients for the recipe described below

The sea monsters in question here are called triops. They're a kind of desert shrimp, which complete their entire life cycle in just 20–40 days. Like horseshoe crabs (which they resemble), triops are living fossils that have basically remained unchanged since the Jurassic era.

Triops are found in pools in deserts that fill up after sudden rains and last only a few weeks before drying up again. Before the triops die, they lay eggs that are fertilized and become cysts. When the pool dries up, the cysts remain in suspended animation. They can survive many years and stand up to extreme conditions. Once they are soaked again by rain, the cysts hatch and the triops go through their life cycle again.

They are, indeed, omnivorous. They'll eat pretty much anything you give them (including insect larvae, and even tenderized plant roots), and they need up to 40 percent of their body mass each day to feed their growth and rapid life cycle. It's just good for us that they only grow to about 3 inches long, or we might be in trouble (of course, if we perfect that enlarging ray, our route to world-domination may be pretty short!).

BASIC STUDY OF A TRIOPS' LIFE CYCLE

This project is a little different from the rest in this book because we're using an over-the-counter kit as part of our materials and procedure. This isn't intended as a cheat, but merely a short-cut to get us to deeper, more interesting science. Plus, if you don't have a desert nearby, it may be rather tough to find any triops on your own in the wild.

STEP 1: First, we need to hatch some triops eggs from our kit. In the small plastic container (rinsed, if it is used), add non-chlorinated (preferably distilled) water, and the eggs and nutrients from your kit. If you want to try raising several generations, only use part of the eggs from the kits. While part of our experiment involves setting up a breeding tank and growing our own future generations from the original triops, we may not succeed on the first go-around, and it'll be handy to keep some eggs in reserve. Plus, only between one and three triops usually survive from any generation (they tend to eat each other—survival of the fittest!).

The temperature of your hatching tank should be 73–85 degrees Fahrenheit (23–30 degrees Celsius). If your laboratory (always say this word in the proper, UK English pronunciation: "lah-boor-uh-tor-ee"—it just sounds more mad-scientisty that way) is colder, cover one end of the tank with aluminum foil for shade and shine a gooseneck lamp with an incandescent bulb (if available) on the tank for warmth. If you've mandated all compact fluorescents in your lab, you could try letting the exhaust air from a laptop computer blow on the container. Point is, you need to maintain the temperature carefully, neither letting it get too cold for them to grow, or so warm that they cook (at least not yet). Put a small room thermometer near the tank to monitor the temperature.

STEP 2: To feed your triops, start by introducing drops of human blood into the tanks once a day. No, not really, though that might be an interesting experiment—will triops feed on human flesh? See if you can get any (un)willing subjects to provide samples!

Actually, within a day or so of rehydrating your triops eggs, you'll see tiny triops swimming around. This early, they can feed on the nutrients included in the kit, so no worries. Since these are basically microscopic organisms, they really don't need much to grow on, even if they are doubling in size every day. But if you need more nutrients for them, you can use a little water from a nearby pond or puddle. There's sure to be plenty of happy microscopic organisms for your triops to eat there!

When the triops are big enough to feed (a few days), you can begin giving them crushed food pellets from the kit, as directed. After a week you can feed the triops bits of carrot and brine shrimp (live or frozen, available from most pet stores that deal in fish) in addition to food pellets.

Excellent, you have raised your first generation of triops to maturity. You granted them life, you control their environment, and you can provide, or take away, their means of existence. You are as a god unto them. Will you be a benevolent or wrathful god? And, really, what now? All this power is great, but don't let it go to your head. The next step is vital for ensuring your ongoing divinity.

From the Mad Scientist's Notebook

There's more to taking care of your triops than just keeping them warm and feeding them each day. You'll also need to change some of their tank water each day, replacing it with clear, distilled water again. But change no more than 25 percent of the tank water on any given day, or the stress will kill the triops.

The Web site www.mytriops.com has pretty much all the information you'll ever need to be a successful triops care-person, so check it out.

STEP 3: Now we're going to set up a breeding tank so you can continue your little (no size jokes, please—it may hurt the triops' little feelings) family.

When your triops get big enough to be crowded in their initial tank, prepare a larger tank. Put a thin layer of clean sand on the bottom of your container and carefully move the triops over (a gentle pour from tank one to two will work). The triops will make interesting patterns in the sand as they trawl for foods. They will also lay eggs in the sand. After a few days in this tank, you can move the live ones to another container, and then let this one dry out completely. Let it sit a few days or more, and then (when you're ready) add water and nutrients just like you were starting from scratch with the kit. You should get a whole new batch from the eggs that were in the sand.

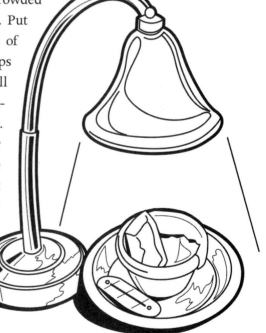

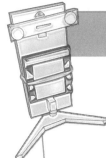

What Else Can I Do?

Under normal conditions, triops double in size every day in the early days, and continue to grow visibly after that. Every few days, they shed their exoskeleton, which you can scoop out and save in a container of water so you can visually track your triops' growth .

Knowing this, you can set up an experiment to see how a variety of different factors can affect their growth and lifespan. Set up two or more tanks with different conditions:

- ▶ 24-hour light v. no artificial light
- ▶ Different temperatures (test the extremes of the viable temperature ranges)
- ▶ Try a variety of nutrients, or differing feeding patterns.
- ▶ Vary the sizes of containers.
- ▶ Introduce new physical conditions into the situation—add rocks to the floor bed or swirl the water on regular intervals to replicate storms.

You can even try training your triops! According to available literature, triops can be trained using food or light.

Hypothesize whether you can induce specific behavior in your triops by conditioning with food rewards. See if you can induce them to move to a specific part of their tank by tapping, and then dropping in food. Do this for a few days, and then make the amount of time between tapping and adding food longer and longer. Can you get to the point when the triops respond to the taps without your needing to add food?

Or study the triops' behavior when lights of different types are shone on the tank, or even parts of the tank. Research their "eyes" and hypothesize what different intensities or colors of lights might do to their behavior.

SEA MONSTERS, OR FOOD OF THE FUTURE?!

One thing we can study with this experiment is whether triops could potentially make a viable protein source for the future (consider this a preview section for Apocalypse Survival Science).

Considering their hardiness and quick life cycle, as well as their omnivorous nature, it's not a crazy idea that growing triops as a food source might be a viable way of ensuring a sustainable, nutrient rich food. Obviously, as with most any food, it's going to be about how you cook it (they are, after all, a shrimp—so treat them as such when cooking). I like Mediterranean fare, so here are a few suggestions:

▶ By themselves, you could add them (after cleaning and slicing into fine strips) to a sauce made by sautéing minced garlic and butter, with a little white wine added.

▶ You could brown the butter, and then add some mizithra cheese and the triops (boiled first, and chopped). Serve over an angel-hair pasta and you'll never know the triops are even there.

▶ Braise the triops in olive oil, then chop them together with some kalamata olives, chop finely, and stuff into mushroom caps with a chunk of bleu cheese on each.

Of course, it may take a special kind of mad scientist to cook and serve creatures you've raised yourself, but we'll leave that up to you. The real challenge is to see whether you can get classmates to try your delicacies when you make your science fair presentation. If you succeed, then you really do deserve to be called a mad scientist!

Exploring Fluid Dynamics
The Magic of Mentos and Soda

In June 2006, Stephen Voltz and Fritz Grobe posted a video on the Web site of their entertainment company Eepybird, entitled "The Extreme Diet Coke and Mentos Experiments: Experiment #137," which demonstrated the effect of dropping Mentos candy into 101 bottles of Diet Coke. It was meant to simulate the fountains at the Bellagio casino in Las Vegas. With that, a viral video sensation was born, and exploding soda became a worldwide phenomenon.

Since then, Stephen and Fritz have performed all over the world, giving a show that mixes equal amounts of giddy, geeky fun and cool science. It's gotten a load of coverage, too, both on TV (*Myth-busters* spent half an episode on it), and even in the *American Journal of Physics*. But it's such a visually stimulating (and downright fun) effect that studying it for a science fair is almost irresistible.

EXPERIMENT	EXPLORING FLUID DYNAMICS: THE MAGIC OF MENTOS AND SODA
CONCEPT	Certain candies and sodas react in almost explosive ways. This project will explore what you can learn by playing with the variables in the reaction, including materials and the release mechanism.
COST	$—$$
DIFFICULTY	⚙
DURATION	☼
DEMONSTRATION OR EXPERIMENT	This project may be conceived as either a demonstration or an experiment. Either way, it's suggested that the practical lab work be done outdoors, and in a place that's easy to clean up.
TOOLS & MATERIALS	• Soda (depending upon your hypothesis, either just diet soda or a mix) • Mentos candy or other candy • PVC pipe and fittings • Glue (Gorilla Glue, or a strong silicon sealant) • Large paperclips or heavy wire • Drill • $1/16$-inch drill bit (or smaller).

Soda is an interesting material. Most sodas are basically flavored water that has been carbonated. What does that mean? Well, you know what carbon dioxide is, right? It's the gas we exhale, and is the result of most forms of animal respiration. It also happens to be soluble in water (though not VERY soluble).

When CO_2 is dissolved in water under pressure it forms a very tenuous relationship with the H_2O. It doesn't take much encouragement for the CO_2 to go from being dissolved back to being a gas again. It just needs a little agitation. Sitting and letting it get warm will do the trick. Eventually most of the CO_2 just turns into little bubbles of gas, floats to the top, and makes its escape into the German countryside on a stolen army motorcycle.

Or it just blows away. Same difference.

What's My Hypothesis?

So, how do we build a science fair experiment around the diet soda and Mentos reaction? Well, we have a few choices, if we consider the variables possible in the reaction:

▸ **Soda Type:** Base your hypothesis on what type of soda you think will react better (more forcefully, longer-lasting reaction). Is diet or sugared the best? Do colas work better, or can the various fruit-flavored varieties spout just as well? Is there a way to set up a control using distilled water and a home carbonation system like Soda Stream?

▸ **Candy Type:** Mentos are the most famous, but do you have an idea why? (more on this in the science section later—spoilers!) Want to experiment with other candies and see if you can find something even better?

▸ **Amount of Candy:** If you're using Mentos, is there an optimal number of candies to drop in? Could altering the candies in some way (break into pieces, grind into powder, roughen the outer surface), improve or ameliorate the reaction?

▸ **Amount of Soda (bottle size):** Since soda bottles of different volumes still have a standard opening size, it's not hard to test various volumes with the same spigot (we'll make our own spigot and remote candy release spigot mechanism that will screw on the top of the bottles). Will a 12-ounce bottle spout just as high as a 2-liter?

▸ **Spigot Size:** Since we can make our own spigot, why not try varying the diameter of the release hole and see if there's an optimal size for generating the maximum vertical spout?

Of course, before we get down to experimentation, we need a spigot. They are available commercially online (Stephen and Fritz sell them at Eepybird.com), but it's also fun and easy to make it ourselves. Plus, it adds to the scientific veracity of the project if you can say you made your experimental platform yourself!

I don't think we'll ever know when someone first dropped a certain type of candy into a bottle of diet soda and received the surprise of, if not a lifetime, then of a very long day (especially if it was winter in Minneapolis). I have to imagine that the effect must have been somewhat known about in certain circles well prior to its popularization on YouTube. But it's one of those things that just doesn't get old. Like talking babies in commercials.

Okay, bad example.

MAKING THE SPIGOT FOR TESTING

STEP 1: Cut yourself a 2-inch piece of $\frac{3}{4}$ inch diameter schedule 40 PVC irrigation pipe.

STEP 2: Take the cap from a soda bottle and drill a $\frac{3}{4}$-inch hole dead-center in the top. This may be the most challenging part. You might want to make a starter hole with a smaller bit, and then finish it with the larger one.

STEP 3: Using Gorilla Glue or a similar adhesive, attach the top of the cap to one end of your PVC. Prepare the pieces per the instructions, apply the glue, and use a clamp to keep the pieces attached for the recommended drying time. We need a good seal that will hold up under pressure.

STEP 4: Take a $\frac{3}{4}$-inch end cap and drill a hole in the end.

STEP 5: Glue the cap to the other end of your PVC using standard PVC glue or a silicon adhesive (Gorilla Glue will not be quite right for this).

From the Mad Scientist's Notebook

<SPOILER ABOUT THE HOLE IN THE CAP> A ¼ inch diameter hole works very well for generating a spout, but if height it not an issue for you, or you need to be able to measure the height of the spout from an accessible position, you may want a larger diameter hole, which will make for a shorter spout.</SPOILER>

STEP 6: About $\frac{1}{4}$ inch up from the soda cap, drill a $\frac{1}{16}$-inch diameter hole directly through one side of the PVC and out the other.

STEP 7: Straighten out the large paperclip, then put a little loop in one end and, if you like, tie some string or ribbon to it to give you the ability to trigger the candy drop from a safe distance.

Okay, there, you're done! You now have your testing spigot. Of course, just for verification, we really ought to, uh, test it once, right?

BASIC PROCEDURE FOR A SODA FOUNTAIN

First, make sure you're doing this in a location where a spray of soda won't make a mess that someone might object to; perhaps a back patio that can be washed down easily with water.

Hold your spigot upside down and put three to five Mentos candies inside the tube. Insert the paperclip through the tiny side, hold on one side, and back out the other so it keeps the candy in place.

Take a 2-liter bottle of diet soda (Coke, Pepsi, RC Cola, or otherwise) and open it, removing the cap. Replace the cap with your spigot, tightening it enough to make sure there will be no leakage, but not so much that you break the seal between the spigot cap and the PVC.

Set the bottle carefully on the ground. Take the string/ribbon from the trigger and walk back as far as you can.

Count down, and pull.

You should hear and see the candy fall in. There will be a brief pause, and then the soda will erupt in a geyser up to 20 feet high. There may be some leakage out from the trigger holes, but it will be negligible.

What's the Science Here?

Which study that has gone into this phenomenon suggests that the diet soda and Mentos reaction is a perfect storm of materials and conditions that causes the rapid release of CO_2 out of the flavored and colored water mixture? It appears that the aspartame in diet soda creates a lower-than-normal surface tension for the liquid. The gum arabic (a gummy substance commonly used in foods) coating of the candies, their rough exteriors, and their density cause them to sink quickly and start over-encouraging the bubble release of CO_2 from the solution (it was already only weakly dissolved). Smaller bubbles coming up from the bottom generate more and bigger bubbles as they come up, and it all snowballs in a rush to the top, where the very small escape hole helps increase the pressure so that the foam shoots out.

If you want the details, there is an article available from the *American Journal of Physics* (*American Journal of Physics*—June 2008—Volume 76, Issue 6, p. 551—http://dx.doi.org/10.1119/1.2888546) that really digs into the nuts and bolts.

EXPERIMENTATION OR DEMONSTRATION

This is a very visual reaction, and can make for an excellent demonstration. You can even pull it off indoors. For the 2011 Maker Faire, we built a clear plastic column out of a long roll of clear 5-mile acetate, supported it over a bucket, and set the soda off right in the middle of our booth to great effect.

If you're doing experiments, the biggest challenge is measuring fountain heights, since the reaction happens so quickly and you're often darting around pulling the trigger and running away from the splashes. A good strategy to deal with this is to set up a measuring reference behind your test area and record the tests on video.

If you purchased PVC for the spigot, maybe you have an extra couple of 8-foot lengths lying around. Connect them with a coupler. Measure and mark inches and feet with colored markers, and then set the pipe up vertically just behind your test area, maybe by tying it off to the branches of a tree, or using clamps to hold it against the eaves of your house or the cross-beam of a play set.

Set up a video camera on the other side of your test area, far enough away so it won't get wet, and angle it so it can record all of the vertical measuring pipe. If you can get one that shoots 60 frames per second or faster, it'll be even better, since you can slow down the playback to better see how high the peak fountains happen.

Start the video camera, and as you run each test, clearly present the information about the test to the camera—explain what variables are staying the same, and what variable is changing. Include date, time, and even weather conditions. It's all important scientific data!

When you're done, you can take the video to your computer, watch it, and transcribe your data for final analysis. After you've cleaned up the patio, of course!

ANALYSIS AND RESULTS

So, what should you expect from these experiments? Well, that all depends upon what you were testing.

If you tested different soda types using the same candy (preferably Mentos), you may not find much differentiation between the brands or types. However, the popularly held belief is that the sugar-substitutes in many diet sodas make the reaction stronger than naturally sweetened sodas.

If you tested different candies, you may have much to analyze. Can you figure out (based on the science explained above) why candies other than Mentos may have similar or different results?

And always track external factors that may impact results. What kind of day was it when you tested? Warm or cool? Did you think about trying the same experiment, but with the soda at different temperatures?

Because of the visual nature of this experiment, the graphics you may reproduce from the videos you took will be great for a presentation or science fair board, but always back them up with well-formatted data.

Of course, if you can get your school to let you launch a few soda fountains at the fair, you may not have a problem winning "most popular experiment"!

Understanding Calories

Junk Food in Flames

Idea by Chuck Lawton

The mad scientist is really the champion of the people. We see that society is corrupt, that the populace is being misled by politicians and corporations all vying for your attention and money by telling and selling you what you want, not what you need. A good mad scientist wants to take over the world in order to show everyone the truth.

Especially the truth about the food we eat.

And while we could put our trust in the information on the food labels that come on everything we buy at the supermarket, as aspiring mad scientists, sometimes we want to do it ourselves. Indeed, it is possible to determine the calories in the food we eat ourselves at home, at least to some limited degree of accuracy. And the best part? We get to burn things!

EXPERIMENT	UNDERSTANDING CALORIES: JUNK FOOD IN FLAMES
CONCEPT	You can measure the calorie content of foods by burning them and measuring the change in temperature of a known volume of water.
COST	$
DIFFICULTY	⚙ ⚙ — ⚙ ⚙ ⚙
DURATION	☀ ☀
DEMONSTRATION OR EXPERIMENT	This project is a demonstration of the concepts, which can lead to experimentation.
TOOLS & MATERIALS	• 2 12-oz. soda cans • One old-school glass-tube thermometer • Large paperclips • Measuring cup • Box cutter for cutting the soda can • Can opener for taking the top off a soda can • Junk food or nuts for experiment • Candle lighter or matches • Fire extinguisher • Safety goggles

You can, indeed, take the science of determining the caloric content of foods into your own hands (at least roughly). This project involves building your own calorimeter—a device used to determine the caloric content of food by burning it and measuring the amount of energy released in the combustion.

The design of our primitive calorimeter uses two soda cans stacked on top of each other. The bottom can will be the burning chamber, and the upper can will be the water reservoir where we'll monitor the temperature change. Energy released from the burning food will heat a 100-ml. sample of water in the upper can. We take temperature readings in the water before starting the fire, and just after it burns itself out. The difference in start and finish temperatures is used to calculate the amount of heat energy the food gave off.

What's the Science Here?

There are two types of calories that we'll be exploring in this experiment. *Big C* Calories (note the uppercase C) are typically used to describe the amount of energy present in food, and are also known as dietary calories. *Small c* calories (they use the lowercase c), also known as gram calories, describe the energy required to raise the temperature of 1 gram of water by 1 degree Celsius. In this experiment, we will calculate the number of calories present in some junk food and convert that number to the dietary calories you are all familiar with.

And please note: We're going to work in metric units for this experiment. This kind of science really demonstrates why METRIC UNITS ARE SO MUCH BETTER THAN IMPERIAL UNITS. The complexity of interchanging the conversions for British Thermal Units (BTUs) and liquid (mass) ounces of water versus volumetric ounces of water would be a nightmare. Instead, much of this math works on the very simple principle that 1 ml of water = 1 gram of water. And as mentioned above, 1 calorie (small c) is the amount of energy required to raise one gram of water by 1 degree Celsius. SO SIMPLE!

This will not be a perfect experiment. I can promise you that, even were you to do everything just right to set up this design, your results would not match up with the information on the bag's nutrition label. Under ideal circumstances, all of the heat generated from burning the food is transferred to the water. But as you can see from our apparatus, there is plenty of opportunity for heat to escape out the sides and bottom of the can. Real, professional labs that test food for this data have calorimeters with extensively insulated burning chambers, and highly sensitive equipment that captures all the heat released. We've just got a couple of soda cans and a thermometer.

But . . . that's okay.

Part of this experiment involves understanding imperfections in our calorimeter and comparing the results to the published calories on the food you're burning. The difference is called the *% error* or *% yield* of the experiment. And knowing the science around this procedure, one very good possibility for a science fair experiment you could develop from this project is to design your own, more efficient calorimeter.

From the Mad Scientist's Notebook

Part of the challenge of this project is picking foods to test that will give us the best results. To achieve the best results, we need a couple of things to happen: The food needs to light on fire easily, so that our attempts to light it don't transfer measurable heat to the water above; and the food needs to burn as completely as possible, so that our results correlate to the values for a whole piece.

Junk foods like Cheetos, Fritos, or other processed snacks, or foods high in oils like nuts, work best, as the oils and fats in these foods burn easily. You want to avoid anything that has a high moisture content, like fruits or vegetables.

Important: This is the kind of project where you'll need a safe, durable work surface, and a well-ventilated space to work in. There will be fire and smoke from this experiment. If you can, it's preferable to work outdoors on a day without wind. If you are performing this experiment indoors, choose an area with plenty of windows and a ceramic or glass surface. You could also do it on a stove with a vent hood with the fan running. But have a fire extinguisher handy in case the fire gets out of hand (really, it shouldn't—at most, you'll get flames licking the outside of the test chambers; it'll be cool, just don't freak out).

BUILDING THE CALORIMETER

STEP 1: Thoroughly clean out the soda cans, to make sure there's nothing left inside. Pick one can to use as the ignition chamber (aka the place we'll burn stuff).

STEP 2: Use a can opener to remove the top of this can (interestingly, this will work pretty much the same as opening canned vegetables— bet you never thought that the top of a soda can was so similar to other cans!).

STEP 3: Using the box cutter, cut a rectangular hole in the side of the can around $1\frac{1}{2}$ inches wide and nearly the entire height of the can. BE CAREFUL as the edges of the cut can (as well as the blade of the box cutter) will be extremely sharp!

STEP 4: Unfold a large paperclip until it's straight, and then re-bend it to a tall shape, where the top and bottom legs are about $1\frac{1}{2}$ inches long, and face the same way. Punch a small hole with the paperclip, and insert one end into the side ignition chamber, about 1 inch from the top of the can, piercing the aluminum. Insert enough of the paperclip to allow your food source to sit in the middle of the can. Adjust the bottom leg so it can sit under the can, using the weight of the construct to keep it anchored so the food, when planed on the top arm, will stay still.

STEP 5: Measure 100 ml. of water using a measuring cup. Many Pyrex or other glass 2- and 4-cup measuring cups are labeled in both imperial units and metric units.

Note: If you cannot measure 100 ml. use $\frac{1}{2}$ cup of water instead. In the calculations later in this project, substitute 118 ml. as the known amount of water.

Pour the water into the uncut can. Insert the glass tube thermometer into the mouth of the can. This upper can is our crude calorimeter.

STEP 6: Measure the starting temperature of the water in the upper chamber of your calorimeter in degrees Celsius with your thermometer. Estimate to the nearest tenth of a degree, if you can, for accuracy.

STEP 7: Be sure you are in a safe, well-ventilated space. Using a candle lighter or a long match, light the food that is suspended from the paperclip. When the food is able to hold its own flame, withdraw the lighter or match, to keep the extra fire from adding heat to the reaction.

Depending on the oil content, the food will flame up rapidly and sustain for a little more than a minute. Flames will likely escape the ignition chamber and lick the sides of the calorimeter.

STEP 8: As soon as the fire extinguishes itself, take another measurement of the water temperature in the calorimeter. Again, estimate to the nearest tenth of a degree Celsius.

You'll want to perform the experiment multiple times to obtain as much data as possible, help overcome one-time errors, and normalize the data for an "average" sample of your food, since each one will be a little different. Make sure you clean the cans each time, and let them cool down to room temperature. Also, replace the water each time, because you'll lose a little water during each experiment due to evaporation. You have to have 100 ml. of water to start with every time you perform the experiment.

CALCULATING THE CALORIES (AND THE CALORIES)

For each sample, calculate the difference between the final temperature and the starting temperature. For example, if you burned three average-size Cheetos, your data might look as follows:

OBSERVATION	STARTING TEMP	ENDING TEMP	DIFFERENCE
1	22.6	33.2	10.6
2	22.8	31.6	8.8
3	21.8	31.4	9.6

If you were only able to record your temperature data in Fahrenheit, use this formula to convert your temperature data to Celsius:

$$\text{Temperature}_{Celsius} = (5/9) \times (\text{Temperature}_{Fahrenheit} - 32).$$

Next, you'll want to average all of your difference values by adding them all together, then dividing by the number of observations [for the data above, (10.6 + 8.8 + 9.6)/3 = 9.6 as the average of the three results]. The average value is important because we are looking for the number of calories in an "average" sample. Just looking at most types of food, from chips to nuts, no two are identical, but if we study a large number of them and find the average, the data we've taken and the results we calculate are a valid approximation of any given sample you'll ever pull out of a bag or can. It also helps to address any inconsistencies between one run and the next. Over a large enough sample size, any statistical errors or outliers affect the results less.

Now you're ready to calculate calories. The formula for this conversion is:

$$\text{calories} = \text{mass } H^2O \times \Delta H^2O \text{ °C} \times 1.0 \text{ cal/g/°C}$$

Let me translate. The number of calories (representing the amount of energy being generated by the burning of your sample) is equal to the mass of water in the calorimeter, multiplied by the delta, or change in, temperature in degrees Celsius, times the (very simple) constant that states it takes 1 calorie per gram of water to change the temperature 1 degree Celsius (sound familiar?).

The mass of H_2O is easy: 1 ml. of water = 1 gram. Because we used 100 ml., your mass is 100 grams. The ΔH_2O °C is the average change in temperature in the calorimeter, in degrees Celsius—the very average value you calculated three paragraphs ago. The last bit makes the units all work out right.

In the example data above, the formula would be:

$$\text{calories} =$$
$$100g \text{ (mass of the water)} \times 9.66 \text{ (average change in water temperature)} \times 1.0 \text{ cal/g/degC}$$

This equation results in:

$$\text{calories} = 966$$

Now those are lower-case c calories, which aren't the ones listed under nutritional information. Actually, technically speaking, the *Big-C* Calories shown on nutritional information labels are really kilo-calories, using the metric prefix kilo, which means 1,000. That is, each *Big-C* Calorie is equal to 1,000 little-c calories. As such, the sample data result shows a result of 0.966 Calories in a single, average Cheeto sample.

Okay, that sounds reasonable. And remember, this has been done already by reputable laboratories, so we can validate our result by looking at the actual information on the nutrition label on the bag.

For the example above, the Cheetos came from a "Big Grab" size bag, which shows them as having 320 calories in the serving of 1 bag. There are about 80 average-size Cheetos and a handful of crumbs in such a bag. 320 Calories divided by 80 Cheetos equals around 4 Calories per Cheeto. Our calculations derived roughly 1 Calorie per Cheeto. Our data is kind of off.

But that's okay!

This discrepancy can be attributed to the inefficiency of our crude calorimeter and our ignition chamber's inability to transfer the entirety of heat to the calorimeter. Remember all those flames licking out of the ignition chamber? That's lost heat that didn't do anything to the water in the calorimeter. Plus, the aluminum in the two cans got pretty warm as well. All those losses affected the results. And while that's a problem, and one we might be able to solve by improving the design of our test equipment (a challenge I'll leave up to you), what's more important for us (and every scientist out there) is that we can explain it and take it into account.

Calculating the difference between the observed and actual values is the *percent yield*, which is calculated as the observed value / true value. In this case, our apparatus yielded $\frac{1}{4}$ or 25 percent of the true value. And if you know this, then you can use that factor when testing other items for which you don't already know the "official" results.

So, to make this into a science fair experiment, you have a LOT of room to play. First, you could attempt to improve the design of your own calorimeter using different materials and finding a way to make sure more heat gets transferred to the water (perhaps a larger combustion chamber to keep the flames on the inside, or some kind of insulation to limit heat loss). You could test your design using the same material, for which you know the lab-tested results, and see how the results you get compare as you adjust and tune it. Or you can calibrate your calorimeter against a known result and figure out the % yield for it, much as we did above, and then use it to study other foods for which you do not know the answer.

Just don't eat all the extra, unused samples from the junk food you test, or you may never get out of your lab again.

Thermodynamics
Keeping It Cool without Electricity

Idea by Kathy Ceceri

One of the key features of the mad scientist's kitchen is the two-story walk-in refrigerator. You can keep anything cold in there! The only drawback is the power bill; the local utility is threatening to shut us down. We've tried a number of our own solutions to keep down electricity costs, but we haven't quite been able to keep the giant hamsters (the ones we genetically engineered to run in the giant wheels to spin the turbines) from exploding every forty-seven miles. So we need to figure something out, because what mad scientist *doesn't* need their cup of ice cream before bedtime each night?

Obviously, if we're going to have our ice cream, we're going to need to keep it cold, and that means we need to devise some kind of cooling system that will run without electricity. At least until we get the hamsters sorted out.

EXPERIMENT	KEEPING IT COOL WITHOUT ELECTRICITY
CONCEPT	Use the Second Law of Thermodynamics to design a simple evaporative cooler that keeps foods fresh longer without electricity.
COST	$
DIFFICULTY	⚙
DURATION	☼ ☼
DEMONSTRATION OR EXPERIMENT	Either; can form a hypothesis about which design will work best, then test and measure it.
TOOLS & MATERIALS	• Butter • Plastic wrap • Digital food thermometer • Terra-cotta flower pots • Terra-cotta flower pot dishes • Disposable bowls and cups • Sand • Cloth (we used a bandanna) • Water

Refrigerators work because of the Second Law of Thermodynamics, originally conceived by Nicolas Carnot in 1824, and more clearly stated by Rudolf Clausius in 1850. The law basically says that heat energy degrades over time—a process called entropy. Heat always flows to an area of lower heat, never the reverse (unless you add more energy to the system). This has a whole bunch of ramifications with respect to how machinery works, and explains why the fabled *perpetual motion* device can never actually exist. But for us, the law is important because it helps us keep our ice cream cold!

When a liquid turns to a gas by evaporation, the energy that was used floats away with the evaporated liquid, taking heat with it. In the 1750s, scientists like Benjamin Franklin experimented with cooling things by evaporation. They used liquids that evaporate quickly, such as alcohol. But these "volatile" liquids are also dangerous to handle.

However, even evaporating water can keep a container cool. In West Africa the *zeer pot* is a traditional kind of evaporative cooler that has been updated for wide use where electricity is unavailable. In Australia in the early twentieth century, housewives used evaporative coolers when they didn't have electric refrigerators. In many places where a hot climate is the norm, devices called *swamp coolers* are used as a less expensive alternative to traditional air conditioners. Swamp coolers are evaporative coolers, simply using fans to blow warm, dry air through pads that are kept constantly wet. The incoming heat evaporates the water, dissipating heat energy and causing the exhaust air to be cooler (and moister) as a result.

For this experiment, we will use the same principles to keep a small space cold enough to store ice cream. The process outlined below will help you think about potential alternative designs that you could come up with yourself, and perhaps include in a science fair experiment.

STEP 1: Design Your Evaporative Cooler

To design your cooler, there are a few important criteria to consider:

1. Evaporative coolers consist of containers that can soak up water and let it evaporate into the atmosphere, thus expending heat energy and becoming colder in the process. They work best in hot, dry climates, where water evaporates quickly. It is useful to position them in direct sunlight and in a space with good airflow to keep the evaporation happening as actively and constantly as possible.

2. Because the evaporation carries the water away from the soaked container, water must be added back into the system on a regular basis. So the best designs include a supply of water that can replace the water that has evaporated.

3. Cost of materials is an important consideration, since the cheaper and more plentiful the parts are, the easier they are to replicate for the most people.

From the Mad Scientist's Notebook

Here are some possible designs you can try out (or skip this section if you want to experiment with designs on your own). They are based on real-life evaporative coolers used throughout the world and will give you an idea as to how the system should function. Use these as your baseline for studying the concepts, then formulate and build your own designs.

▶ Traditional coolers are usually made from terra-cotta clay pots, the same material that flower pots are made out of. These clay pots can soak up a lot of water. One typical design has two pots, one inside the other, with a layer of sand, wool, or other absorbent material in between. The outside pot is soaked with water, and so is the sand between the two pots. As the outer pot dries, it pulls more water from the sand layer toward it through osmosis.

▶ Another common design uses a pot turned upside down in a dish of water. A cloth over the pot wicks water from the dish, keeping the pot wet.

STEP 2: Prepare the Experiment

The bottom line of all this work is to preserve food by keeping it cold. That means we need to pick some kind of food to be our testing material inside our cooler designs. And while we are ultimately hoping to keep our mad scientist ice cream cold, we're going to use another material for testing (why would we want to waste all that perfectly good ice cream?).

Instead, butter makes an excellent food to test an evaporative cooler with. It is easy to measure the temperature of butter, and it was actually the food for which these kinds of coolers were used in the early twentieth century all over America and the UK.

Depending upon how many designs you're testing, leave one or more sticks of butter out at room temperature long enough until it is soft, but not runny (obviously what "room temperature" is varies from house to house, so keep an eye on it).

To keep the butter clean, put each testing sample into a small disposable cup and cover it with plastic wrap. Use the food thermometer to poke a hole through the plastic wrap and let the thermometer sit in the butter and measure its temperature (unless your thermometer is an instant-read style). Fill one cup for each design you are going to test, plus one to sit out without cooling (your first control), and one more to put in the electric refrigerator (your second control).

STEP 3: Test the Coolers

Assemble your test coolers, put a cup of butter inside each one, and set up your controls. Prepare your datasheet for each sample, as well as the controls, with a description of the given design, how much water your system can hold in one fill-up (not applicable for the controls), and with enough lines to record temperature readings at regular intervals for as long as you plan to carry out the test. Depending upon the complexity of your designs (especially if they have water reservoirs), this could be many days.

When you're ready to go, take a start temperature reading for each sample and start your engines (or coolers, as it were). Set a timer to measure the temperature of all the test cups of butter at regular intervals, but over the course of the experiment, try as much as possible to avoid disturbing the coolers or moving the cups, to allow for as little contamination as possible.

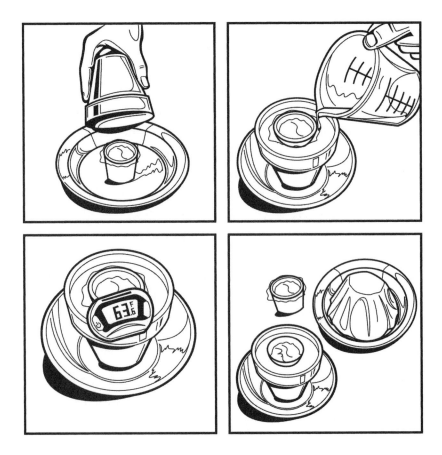

Of course, an important question is, when is the experiment done? Depending upon the longevity of your designs, this could be hours, days, or more. One big indicator that you've exhausted the single fill-up capacity of a given design is that it dries out, which means there's no more active cooling going on. You could also decide that a given design test is over when the temperature of the sample inside reaches within a few degrees of the room temperature. Or when the butter starts spoiling—but that might take quite a while if you're a good designer!

STEP 4: Analyze the Results

This is the kind of data that just cries out to be graphed. Create a graph with temperature on the y-axis and time on the x-axis. Plot the data for each design in a different color, and look at the results. You should have one nearly straight horizontal line at room temperature for your un-chilled control, and another at the temperature your electric fridge is set at for that control. And then you'll have all sorts of interesting curves for your cooler design test.

Looking at the data, you should be able to determine the following for the results of your experiment:

▶ Which version cools the butter the fastest?

▶ How much colder does each design keep the butter, compared to the butter at room temperature?

▶ Which design worked the best overall (and what does that mean)?

▶ What other improvements could you make?

This experiment makes for an excellent science fair project—because of the variations that are possible to study, the ingenuity that's easy to add to personal designs, and the important scientific principles at work. And of course, because it involves food, it always makes a mad scientist happy.

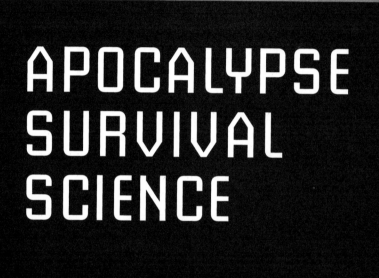

APOCALYPSE SURVIVAL SCIENCE

Building a MacGyver Radio

Idea by Kathy Ceceri

T he aliens have attacked. The robots have turned against us. The zombie plague has, shall we say, "gotten out of hand." In short, the apocalypse has come, and humans have been pushed to the brink of extinction.

But we're fighting back.

Sure, the EMPs (that's electromagnetic pulses) have fried most of our home electronics, but we know the resistance has captured some communications infrastructure, and they are now broadcasting news and instructions about how we can organize the counteroffensive.

Now, if only we had a way to hear those messages!

EXPERIMENT	BUILDING A MACGYVER RADIO
CONCEPT	Use simple materials to detect electromagnetic radiation by converting it to sound waves.
COST	$
DIFFICULTY	⚙ ⚙
DURATION	☼ ☼
DEMONSTRATION OR EXPERIMENT	Easier as a demonstration of the simple technology. However, experiments could be developed around an attempt to find the best materials, or number of available signals.
TOOLS & MATERIALS	• Cardboard base (you can also use a piece of wood) • Brass paper fasteners/brads (for a wood base, use brass thumbtacks) • 100-foot thin insulated wire ("magnet wire" is one handy version) RadioShack Part # 278-1345 • Cardboard toilet paper tube (or round oatmeal box or plastic bottle) • Razor blade (careful!) • Safety pin • Pencil stub, sharpened • Piezoelectric earphone (available from scitoys.com for about $7) • Tape (transparent tape and/or electrical tape) • Paper towel tube • Scrap paper • Aluminum foil • White glue

In this age of microelectronics, all the little parts that make up our slim and sleek electronic gadgets—the resistors and diodes and wires and capacitors and all—live on tiny printed circuit boards. Everything is so small and compact that, if we open one up, we have no idea what the parts really are, and no idea how we could possibly replicate the function of each part ourselves. They are a kind of future tech that, were we to want to, we could never reproduce by ourselves.

"Any sufficiently advanced technology is indistinguishable from magic," said Sir Arthur C. Clarke. We are in danger of losing ourselves to the magic already.

But it has not always been that way with electronics. And the best example of an electronic device that you can still build yourself is one that is still incredibly useful: the radio.

Here's a mind-boggling fact for everyone who has ever had their iPod or laptop run out of juice: The simplest kind of radio receivers don't need batteries. They don't need to be plugged in or charged up, and they don't need the sun or a hamster in a wheel to provide power. They work by converting the electromagnetic energy of radio waves directly into sound waves.

Since the first radio receivers were developed in the 1800s, many different designs and materials have been used. During World War II and Vietnam, soldiers often built *foxhole radios* or *POW radios* using their government-issue razor blades or bits of coal coke as a diode, along with scrounged wire and other used parts.

What's the Science Here?

There's energy all around us. We just have no way to sense or tap it (unless we get a metallic tooth filling that does something funny enough to be on a '70s sitcom). This energy is in the form of electromagnetic waves. There's plenty of it in the background of our world and universe, but there's a lot more being broadcast at us in the form of radio. Radio signals are just electromagnetic waves that have been modulated to carry information. It's possible to collect those waves and use the energy in them to create little bits of movement. That movement can cause pieces of specially arranged metal to do something funny: make sounds according to the modulation of the waves.

We are going to build a simple radio receiver, basically a tuned electrical circuit, that uses a long antenna to capture and use the energy from the AM spectrum of radio waves (AM = Amplitude Modulation, one of the two ways radio waves are manipulated to carry signals), broadcast by radio transmitters, to create vibrations in a flexible surface. According to the modulation of the radio waves, the flexible surface will vibrate the air around it, which is a somewhat complicated way of saying it will make sound. Recognizable sound.

Note: It's important that you have a strong AM radio station transmitting within about fifty miles of your location. This will be much easier in urban and suburban areas.

Here are the major components that make up the radio. We're going to build these from scratch:

- **Tuned Circuit:** Allows us to single out specific frequencies from the radio waves. It has two parts:

  - **Inductor:** A coil of wire.

  - **Variable Capacitor:** Layers of foil that, when adjusted, allow us to tune in different stations with a narrow range.

- **Diode:** An electronic component that only allows parts of the electricity to pass through it.

- **Earphone:** Converts the electricity we've collected and filtered out into sound waves (the "flexible surface" we mentioned earlier).

- **Antenna:** Collects the radio waves.

- **Ground:** Acts like the sink drain for the electricity, allowing it to flow continuously through our circuit and not get all backed-up.

I'll take you step-by-step through building each component, and then through mounting them on a base (either a piece of wood or cardboard, around 1 foot square), hooking everything up, and troubleshooting a potentially fidgety device. Ready to go? Okay!

STEP 1: Build the Inductor

Different radio stations broadcast on different frequencies of radio waves (the frequency is how fast the wave goes up and down). What we're going to do is design a circuit that picks up one specific frequency of radio wave. But we want to be able to adjust what that frequency is just a bit to hopefully pick up a clear signal, hence the tuned circuit.

All the inductor really is, is a length of wire; in this case, a few hundred inches of wire. Now, it might seem impractical to have to build a radio by taking up a large part of your workroom with a 300-inch-long length of wire, but luckily the wire doesn't have to be straight. In fact, to make it really easy, it can be in a coil, wrapped around a cylinder. Gee, I wonder what you have in your house that would make a good cylinder base to a coil?

Why not go take a bathroom break and then get back with your answer?

Okay, yes, the cardboard tube from a toilet-paper roll works really well for this (hopefully you don't buy the tubeless kind—it may be greener, but it won't help us here!).

We're going to wrap the wire around the tube, but we're not going to do it all willy-nilly. Yes, "willy-nilly" is an important scientific term that you should know and remember.

The wire you're using is probably already on a spool (which makes this job seems a little redundant, but that's science for you!), so pull out about 10 inches to start with. Leave yourself a 6-inch lead, then tape the wire to the tube close to one end (it should sort of look like this:——OOOOOO).

Start winding the wire around the tube so that each loop is right next to the last, moving around and around all the way down the tube, as if you were trying to make the perfect mummy. It may take upward of 200 loops to work your way down the loop, depending upon the thickness of your wire.

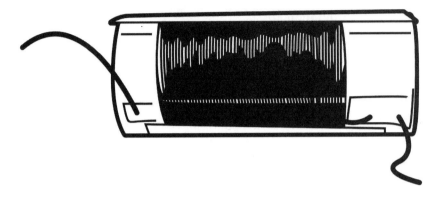

When you get to the other end of the tube where you're an equal distance from the edge as you were when you started, tape it down, then give yourself another 6-inch lead, and snip the wire off. You can also secure a piece of tape across the loops of wire to make sure they don't slip off.

From the Mad Scientist's Notebook

MacGyver tip: Instead of a toilet paper or paper towel tube, try wrapping the wire around a cardboard oatmeal or salt box, or a stiff plastic bottle—anything non-conductive. Indeed, testing different cores for your inductor to determine the best material could be your science fair project.

STEP 2: Build the Tuner (aka a variable capacitor)

This is actually an optional component, and if you hooked everything else included in this design up right now, you'd probably be able to hear . . . something. Maybe static, maybe a single nearby station. But we want more, right? I mean, the resistance may have to change frequencies to keep the robots from discovering them and quelling our nascent rebellion!

So, in simple terms, we need a tuner. Though there are a variety of ways to do it (check the references at the end of this project to see some other designs), for our purposes here, we'll use a variable capacitor. A variable capacitor will allow us to change how much electrical current is running through the radio circuit via the manual adjustment of two conductive surfaces combined to create a place for electricity to take a brief time-out.

The capacitor stores electrical charge in two pieces of conductive material (we'll use aluminum foil) separated by an insulating piece of material (we'll use paper). By adjusting how much the conductors overlap each other, we can adjust the energy storage and tune the radio. We're going to make this an easy task by wrapping foil around a cardboard tube, then making a slightly larger-diameter tube of layered foil-and-paper, that will slide over the outside the first tube, kind of like putting wrapping paper around a rolled-up poster (but not sealing the ends). We will be able to slide the outer tube up and down on the outside of the inner tube, which will change the storage capacity, adjust the frequency being absorbed, and tune the radio.

So, to do this, take a long cardboard tube (paper towel tube works really well) and wrap aluminum foil around roughly half of it (but leaving about an inch of exposed cardboard at the bottom). Attach with a couple of small pieces of tape, but don't significantly obscure the foil.

Cut a piece of paper just a bit bigger than the first piece of foil. Cut a second piece of foil, smaller than the paper (thusly about the same size as the first piece of foil). Roll these together into a tube with the foil on the outside, and size it to fit over the first foil-covered cardboard tube. You want the paper/foil tube to slide over the cardboard/foil tube easily, without tearing the foil, but be tight

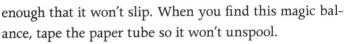

enough that it won't slip. When you find this magic balance, tape the paper tube so it won't unspool.

Now we need to attach wire leads to this so we'll be able to hook it up to the larger circuit. Tape one lead to the foil on the outer tube, and the other lead to the foil of the inner tube (remember, always strip the insulation off at least $\frac{1}{2}$ inch of the wire so when they're taped down, there is metal-on-metal contact, allowing for electrical conductivity).

It's probably smart to have the outer wire as high up on the outer tube as possible, and facing to one side (say left), and the other wire as low on the inner tube (so as to not block any of the sliding action), and trailing off to the other side (say right). Make sure your leads will be long enough to connect to either edge of your final board— you don't want to come up short!

STEP 3: Build the Diode *(the thing that turns electrical energy into vibrations)*

A diode is an electronics component that only allows electricity to flow in one direction. Usually they are pea-shaped little doo-hickies (another technical term), with a wire coming out of each end. Of course, ours won't be quite that elegant, but it'll do the job.

And what is that job, you ask? Well, it's all about the peaks. The diode can be tuned so that it only allows the peaks of energy created by the wave-induced current to pass through. That means on the other side of the diode, there are little bursts of current of different intensity. When those get sent into something like a (very tiny) speaker (or earphone, in our case), they will vibrate the diaphragm and generate sound.

The diode we'll make here creates that effect with some simple household parts. You'll make a "cat whisker," which will be a bent-open safety pin stuck into a sharpened pencil-nub, and you will rest that lightly against a specially treated razor blade. The result will be that all the current travels into one side, but only the peak wavelengths pass through to carry on and drive the earphone.

To make the cat whisker, open the safety pin and bend it to a 90-degree angle "L." Take the sharpened pencil and cut it off so there's about 1 inch of paint above the sharpened point, and you can see the cross-section of the wood and graphite on the cut side. Insert the pin into the cut end of the pencil nub, sticking it into the graphite deep enough to firmly stay put.

From the Mad Scientist's Notebook

The radio will work much better if the razor blade has been "blued," or heat-treated. To blue it, use kitchen tongs or pliers that won't conduct heat to their handles and hold the blade to a flame. It could be a gas stove flame, or perhaps a small home welding torch, or a Bunsen burner. The blade will heat up quickly and acquire a blue-black coloring. When it is uniform all over (this may only take a few seconds), you're done. Remove it from the flame and set it aside to cool before moving on.

The parts of the diode will be the first thing we mount to our base plate, so now is the time to get it ready. The base can be a flat piece of wood or even a piece of cardboard, about one foot square. If you're using wood, you'll need brass tacks hammered into the wood to hold things down and act as connection points. If you're using cardboard, you can use those brass brads you find at office supply stores in much the same way; just poke them through the cardboard and spread our their legs to fasten them down.

Attach the razor blade to your base plate near the front, slightly left-of-center with the blade edge facing to the right, using a brad or tack on either side to hold it down flat. You will have wires to tie off to one of the connection points that must touch the blade, so give yourself a little space to start rather than tacking it down tight to the base.

Positioning the cat whisker is a bit trickier. Using pliers, you'll want to bend the other end of the pin (the head, or the part not stuck into the pencil nub) so you can tack it to the base. You want the pin to act like an arm, pointing straight up from the board, then turning at the elbow, and then pointing the pencil nub down at an angle to touch the blade. Additionally, the whole construct should be able to swivel enough from its connection to the base to let you adjust where on the blade the pencil nub touches. This may take a bit of work to achieve.

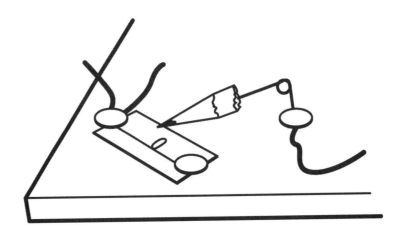

STEP 4: Putting the Components Together

In the end, as with most electronics projects, we're creating a circuit. Electricity comes into a system of components, does a little dance, and goes back out again. In our case, we're sucking the electricity out of the space around us, transforming some of it into the kinetic energy that will drive a diaphragm to generate sound waves in the air, and sending the rest . . . where? Well, to the ground, of course! Really, the earth itself is one big sink for excess electricity, and we're going to add to it.

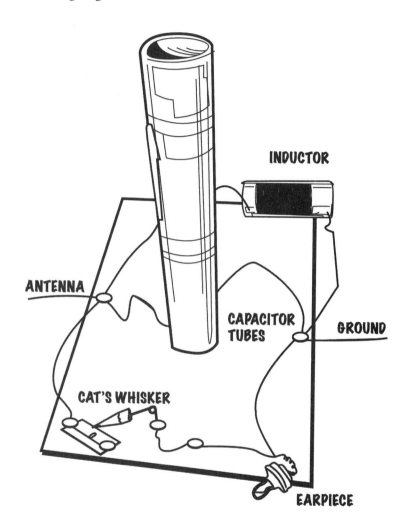

So, we need two brads or tacks on our base dedicated to the receiving and the returning of the electricity. Put one on one side of the board (left), and label it "Antenna." Put one on the other side of the board (right), and label it "ground." We'll come back to them in a bit.

Now you can mount the coil tube to your base plate. It can lie on its side, and you can use the brads or tacks to stick through the bare edges of the tube into the base. Just make sure you don't put your brads or tacks through the wire.

Take the loose ends of wire (the leads) on either end of the coil. Attach one end to the left side "Antenna" connector, and the other end to the right side "Ground" connector.

Stand your variable capacitor/tuner upright and attach to your base plate somewhere near the middle. To attach, use scissors to snip and fold the inch of exposed cardboard you left at the end of the tube into tabs that you can tack down to the base. Take one of the leads and attach it to the "antenna" lead, and the other to the "ground" lead. Sensing a pattern here?

Now cut a piece of wire and strip the ends so you can hook up the connectors you left yourself to a little space on with the razor blade to the "antenna" connector. Once you have the wire twisted onto the razor blade connector, you can tighten it down so the blade won't slip.

Now, to make things simple, we're going to prepare one more thing. Put two more connectors near the front edge of your base, to the right of where your cat whisker is attached, about an inch apart from each other. Cut, strip, and run a wire from the cat whisker to the closer connector. Cut, strip, and run another wire from the other connector to the "ground" connector (which shouldn't be too far away).

Okay, we've got almost everything done. One more component to add, and we'll be ready to go!

STEP 5: Hooking up the Earphone/Speaker

This radio uses only the power of the radio waves, which means it needs a very sensitive type of earphone or speaker to produce a sound. Your best choice is an old-fashioned piezoelectric earphone, which can be ordered online at scitoys.com, or found at your local electronics store. The earphone has two wires. (Be careful, they're very thin and break easily.) Attach one wire to one of the new front connectors you just added, and the other wire to the other connector; it doesn't matter which is which.

From the Mad Scientist's Notebook

For the earphone or speaker, you can also try using computer speakers, a piezoelectric speaker from a "talking" greeting card, piezoelectric buzzer, or the handset from an old telephone (the kind that attaches to the phone with a wire). However, modern earphones (like the ones that come with your iPhone/iPod) probably won't work, since they're used to receiving an electrically amplified signal and won't be sensitive enough for this use.

STEP 6: The Ground and the Antenna

To make a ground, take a longer piece of wire and attach one end to your "ground" connector. Attach the other end to a metal pipe that touches the ground in some way (the real ground, not the floor or some concrete). This could be a variety of plumbing parts in your house. Most any metallic piping that goes somewhere under your house, like your sink drains or water lines, should work. If you have metal radiators, they can work as well. If you can string the wire outside the house to a pipe that obviously sticks into the ground, great. The better the grounding, the better the radio will work.

Depending on how close you are to a strong radio signal, you may need 25–50 feet or more of antenna wire. You can splice shorter pieces of wire together if you need to, by stripping the ends, tying the end of one to the end of another, and taping them up to make the longer wire.

Attach one end to the "antenna" connector, and run the rest to wherever you can. If you can do the project outside, you can try running the antenna along a fence line around the perimeter of your yard. If you're inside an apartment, see if you can get away with simply dangling it out a window. You can also just string it around a room a few times. Just get it as high up off the ground as you can, and make it as long as you can.

STEP 7: Using and Troubleshooting the Radio

Now you're ready to go. Once you get everything hooked up, you may well start to hear something through the earpiece. Maybe static, maybe (if you're lucky), actual radio sounds (i.e., talking or music).

If you get nothing, it's time to troubleshoot. Make sure all your connections are secure and that the metal ends of wires are touching the other metal ends of wires where you have them meeting on a connector.

Move the pencil point of the cat whisker around on the razor blade until you hear static. If you are not hearing any kind of scratching when you move the pencil point, your earphones are not working. Check that the earphone wires are not broken or disconnected. If you hear scratching but not an electrical buzz, you are not picking up the radio signal yet. Keep moving the tip around. If there's still nothing, try re-bluing the blade, or even getting a new blade.

When you hear buzzing, that means you're getting a signal (or the fuzz between signals). Now you can use your tuner to zero in on a station by slowly moving the outer tube up and down.

If the sound is faint, tapping the earphone or the tuner with your finger may make the sound louder. Holding on to the capacitor tube or the antenna may also improve reception, because you yourself can be part of the circuit! Adding more wire to the antenna to make it longer will also make it louder.

If the sound is too low to hear clearly, try putting ear muffs over your ears (being sure that the earphone stays in your ear underneath). Or hold a Styrofoam cup over the ear with the earphone.

Due to a quirk in AM radio waves, you may get better reception and more stations at night.

If you've done everything right, and the stars are aligned just perfectly (well, maybe that doesn't matter), you will have just created an electronic device that doesn't need a visible power source to run. Which is something the robots are going to hate.

Viva la Revolución!

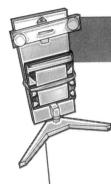

What Else Can I Do?

The following links will give you even more information about how this radio works, and show other variations—particularly for the tuning, and including how to make directly powered versions that up the ante. I'll especially point out the last one, from MAKE, which also has a great video by Bre Pettis.

http://integratedscienceathome.blogspot.com/2011/01/powered-by-radio-waves.html (Kathy Ceceri's original go at this project)

http://www.midnightscience.com/

http://science.hq.nasa.gov/kids/imagers/ems/radio.html

http://peeblesoriginals.com/index.php

http://bizarrelabs.com/foxhole.htm

http://www.ece.cmu.edu/~ee321/spring00/lab3.pdf

http://blog.makezine.com/archive/2007/09/make-a-foxhole-radio-week-1.html

Making Top Secret Invisible Ink

Idea by Jenny Williams

A mad scientist's goal is remaking the world as we know it. One thing that we'll need as we work to build our new society from the ruins of the old is secrecy in our written communications (we can't trust the old phone system—*they* might be listening!). Codes are always good, and geeks like us can certainly work out our fair share of mathematical algorithms and obscure geek culture references to develop an air-tight code, but security works better when it's in layers.

The best way to deliver secret messages is to come up with a way to communicate so that no one else is even aware the communication is taking place. From the beginning of time, people have wanted to send secret messages; whether they were governments with secrets to pass, revolutionaries plotting rebellion, or lovers trading illicit prose. What they used was "invisible ink," basically a substance one could write with (usually via a fountain-style pen, or a brush) that would not show up on the paper until it was specially treated in some way. In this project, it will be our mission to study potential sources for invisible ink, and find out which ones will be the best for our secret network of agents to use in organizing our march toward a new world order.

EXPERIMENT	MAKING TOP SECRET INVISIBLE INK
CONCEPT	Test a variety of liquid substances to determine the best to use for invisible ink. Criteria include how fast the substances dry, how invisible they are when dry, how quickly they are revealed with heat, and how legible they are once revealed.
COST	$
DIFFICULTY	⚙ ⚙ Younger children should do this with close supervision when using substances such as hydrogen peroxide and tools like a hot oven.
DURATION	☀ ☀ The setup and the reveal are fairly quick, but depending on liquids used, the dry time can take a while.
DEMONSTRATION OR EXPERIMENT	This can be a proper experiment when done with untried substances, complete with hypothesis, data gathering, and analysis. You can perform it as a demonstration once you begin using the invisible ink to write messages.
TOOLS & MATERIALS	• Paper strips • Cookie sheet • Oven and oven mitts • Q-tips • Small plastic cups • Stopwatch or other timer. Clear or white liquids to test (Some good options include apple juice, pineapple juice, white grape juice, milk, cream, sugar water, white vinegar, hydrogen peroxide, saliva, lemon juice, laundry detergent, and other sugar-rich or slightly acidic liquids. Bleach is NOT recommended as it tends to ignite in the oven.)

Sending messages using invisible ink is really an age-old tradition. People have been passing secret missives to each other with invisible messages for millennia. What was used for the ink probably depended upon the culture, location, and perhaps even diet of the messengers. Many variations of invisible ink can be made using food items or other household chemicals, so there's plenty to test here.

I'll give you a number of suggestions for materials to try. We're going to stick with materials that write invisibly (or nearly so), and then change color when heated. If you're going to try your own ideas and are using the heating method, first be sure that there is no danger of conflagration for the material.

You can also try to find substances that become visible when treated with another substance through some kind of chemical reaction. Again, be careful and do your research to make sure such reactions are not dangerous. Use safety equipment and be sure to have adult supervision for younger scientists.

STEP 1: Choosing Your Materials

Obviously you have to have materials to test. Most inks are liquids, of course, so we'll start there (although solids like various bar soaps, chalk, or crayons are possibilities for testing as well). Acids, some bases, and liquids high in sugar are good options. Fruit juices of varying types (as long as they'll go on paper relatively clearly), vinegars, clear oils, even un-colored sodas could be good test materials.

STEP 2: Preparing Your Datasheet

Four important factors for assessing the value of an invisible ink include how quickly the ink dries, how well-hidden the dried liquid is on the paper, the time it takes to reveal the message, and the quality and legibility of the revealed message.

You can record this data to different levels of detail, depending upon the rigor of analysis you want to perform. An easy way to structure your analysis is to grade each data point qualitatively on a simple scale:

► **Message Drying:** Quickly, Moderately, Slowly

► **How Hidden:** Invisible, Somewhat Visible, Visible

► **Time to Reveal:** Quickly, Moderately, Slowly

► **Quality/Legibility:** High, Medium, Low

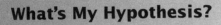

What's My Hypothesis?

This project involves the application of heat to create a chemical change that breaks down larger molecules into components. At least one of the components will have a visible color. This is a process called *thermal degradation*. For acids, carbon compounds break down and the resulting carbon molecules are brownish in color. On the other hand sugars, when heated, go through a process called caramelization. But again, the heat breaks more complex molecules into parts, some of which are brown in color (the parts also have a delightful aroma).

There may be other similar reactions that you can find through your research, and test in your experiment. Based on your research and understanding of heat-activated substances, make a hypothesis about which liquid or type of liquid will make the best invisible ink (acids will turn brown, sugars will caramelize, etc.).

Or, you can record the data in much more detail. Give each of the subjective items a score of 1 to 10 (lowest/worst to highest/best), and actually time the time-based factors. With that kind of data, your analysis can include comparative graphs, and you could even work up a comprehensive scoring system to rank the inks.

Once you've decided on your methodology, create a datasheet to make it easy to fill in the results as they happen in the kitchen—err, "lab."

STEP 3: Setting Up Your Experiments

You'll want one piece of paper, perhaps a 3 inch by 3 inch square, for each testing sample, and you'll need to put an identifying mark on each one so you can ID it through each stage of the experimentation. You may be able to write the substance name in small letters on one edge, or you could create a code for each one (use the same code on your datasheet). Just make sure whatever you write doesn't overlap with your testing area so that it doesn't mix with your sample. This would invalidate your results. And make sure you use the same kind of paper for each experiment, to ensure consistency of the results.

Your datasheet may look like this:

SUBSTANCE	DRY SPEED	HOW HIDDEN	REVEAL SPEED	REVEAL QUALITY
Lemon Juice				
Sugar Water				
Skim Milk				
Cream				
Saliva				
White Vinegar				
Hydrogen Peroxide				
Clear Laundry Detergent				

From the Mad Scientist's Notebook

An important thing to remember when doing any of these experiments: While your goal is to perform your experiments correctly, if you make a mistake, it's not the end of the world. Just make sure you record the error in your results so you can explain what happened in your analysis and what you might change in future revisions of the experiment. Accuracy, detail, and honesty are the keywords in science, and there's truth to the adage that we learn as much or more from our mistakes as from our successes.

Lay the paper test pieces on a clean baking sheet, with plenty of space around each. Prepare your ink materials by putting each of them into a small plastic cup or container that will be easy to access. Make sure the container is clean (no contamination!).

Have a Q-tip-style cotton swab-on-a-stick ready to use for each ink material; that's one each per test substance. Again, we cannot mix the materials when we test them (although, for a more complicated version of this project, you *could* try mixing and testing compounds to see if you can create an even better composite).

You also need to decide exactly what you're going to write with each sample. If you were truly going to use the liquids to write messages, ultimately you'd write them with a fountain pen, quill pen, or with an oriental calligraphy brush. With a cotton swab, we'll be working closest to the brush, so the characters you'll write will be large. Perhaps consider writing one character so you can easily test the legibility. Pick a letter or number and make it something you can write easily without being able to actually see what you're writing (since the ink will be invisible).

STEP 4: Performing Your Experiments

The first stage of testing is evaluating how quickly the materials dry on the writing surface. This is one of those times when it may be helpful to have a lab assistant so you can split up the jobs to make the whole process easier. If you don't have an assistant, you'll have to write the test character and time the drying process yourself one sample at a time. With an assistant, you can write a character and let them record the time to drying. It's probably not a good idea to do two or more samples at one time, as you'll be splitting your attention between them, trying to track start and end times for both, which could get very complicated.

From the Mad Scientist's Notebook

I've made lots of notes in this project about trying to avoid contamination of testing materials. Paying attention to this in any experiment is only going to lead to better results, and perhaps a better grade on a science fair project. For this series of tests, the potential for contamination is high, so you may wish to consider these additional steps:

- ▶ Wear gloves when handling everything, so no oils from your hands can affect the results.
- ▶ If you see any evidence of splashing on your gloves or between samples, immediately replace them with fresh, clean versions.
- ▶ Keep a tidy lab space. After writing each test character, immediately dispose of the swab and sample cup so things become less cluttered over the process of preparing the samples.
- ▶ If you have to move samples around, handle them with a pair of clean tongs, again to ensure nothing from your hands can contaminate them.

When you've written each sample and timed the drying, take a moment to evaluate how well hidden they are. You may want to evaluate them at different angles, with direct and indirect light, and even different qualities of light (sunlight, incandescent light, fluorescent light, candle flame). Record your (admittedly subjective) evaluation of each ink material, and take notes about the positives and negatives of each one (say, if one ink exhibits a particular color, you may later want to try it on a paper that more closely matches that color, or at least suggest that in your analysis).

You may also wish to take photographs of each sample (with the sample ID visible in the image) so you can display the before/after appearance of each on your project board (if you make one).

Now it's time to test the reveals. Preheat your oven to 350 degrees Fahrenheit (ovens vary, but in general this should be a good approximate temperature to work with). As when you recorded the drying times, you may need an assistant here, or stick to only testing a few at a time so your attention isn't too split watching for the appearance of characters. This is an important part of the testing process for you to sort out before you start!

Start the timer, and put a cookie sheet with a set of samples in the oven. Watch it very closely. Different substances will reveal at different rates. The first liquids should reveal themselves in about 20–30 seconds. As they reveal, call out the time to your assistant for recording (or record it yourself), and remove each piece of paper from the tray with a pair of tongs so you can avoid burning them.

Finally, you need to evaluate the legibility of each of your ink test samples. Like you did when you tested how hidden the ink looked, try different lights or angles of light, and make qualitative notes about how well you can read the character you wrote. If you took pictures of the "Before," take the "After" shots so you can compare them. Once you've finished filling in your datasheet, your experiment is finished. Clean up the rest of your materials and tools, after which you can get down to the analysis.

STEP 5: Analyzing Your Results

Now you can analyze. With all this good data, you have plenty of choices of review and presentation. If you used a simple 3-stage criteria (Poor/Moderate/Good), then you can create a grid with full-bubble, half-bubble, empty-bubble graphics (like they use in *Consumer Reports* and other publications). If you have time data, you can plot graphs to give a better visual comparison between the materials. Or you can mix and match how you present the data.

Remember to mention any potential problems, mistakes, or errors you think could have crept in. Could the type of paper have favored one ink material over another? How about the weather (temperature or humidity)? These are all things to take into consideration and mention in your analysis.

Finally, how did the inks do compared to your original hypothesis? Did you find the best possible invisible ink here, or do you need to do more testing? Remember, it's okay to leave a project open-ended, where your conclusion is merely that you need to do more testing. That's the nature of science—very little gets settled in just one try.

The best part of all this is that if you find a truly useful invisible ink material, you can start using it! Send some mysteriously blank notes to your friends to see if they can figure out how to read them.

Warning: Spoilers on the Next Page!

SPOILERS/RESULTS

For anyone doing this project for fun rather than a grade, or if you are a parent or other adult monitoring a student through the process, here are some sample results showing how the datasheet might look after the testing process. This will give you a reference to make sure things haven't gone terribly wrong.

SUBSTANCE	DRY SPEED	HOW HIDDEN	REVEAL SPEED	REVEAL QUALITY
Lemon Juice	Fast (8s)	Poor (3—yellowish)	Moderate (35s)	Good (9)
Sugar Water	Fast (9s)	Moderate (4—grayish)	Moderate (37s)	Good (8)
Skim Milk	Fast (8s)	Moderate (5—yellowish-white	Moderate (41s)	Good (8)
Cream	Fast (8s)	Moderate (5—yellowish-white)	Moderate (38s)	Good (9)
Saliva	Fast (9s)	Good (10—clear)	Moderate (31s)	Moderate (5—mostly an outline)
White Vinegar	Fast (6s)	Good (8—white & shiny)	Slow (66s)	Poor (2—nearly illegible)
Hydrogen Peroxide	Fast (8s)	Good (8—white)	Slow (48s)	Poor (3—hard to read)
Clear Laundry Detergent	Fast (9s)	Moderate (4—Blue-ish)	Slow (76s)	Poor (3—hard to read)

Steam Power, Steampunk Style

Idea by Natania Barron

For an up-and-coming mad scientist in a post-apocalyptic world, it's important to look at what kind of impact you'll make in your endeavors. I'm not just talking about what you'll do. Pulling the last dregs of humanity together into a productive, well-managed proto-industrial society is all well and good, but what kind of aesthetic are you going to develop? Do you want to be the *2001: A Space Odyssey* retro-future kind of mad scientist and create a world with all clean white lines and insane artificial intelligence? Or maybe you're more attracted to a *Tron*-style neon, glow-in-the-dark inside-the-network electronica-fueled rave? It's important to ask yourself these questions.

For me, I envision a future with steampunk. You can't beat its combination of form and function—a science-fiction subgenre and design aesthetic that imagines a world built upon Victorian-era team technology, combining that civilized era with the beauty of its intricately built machines made with bronze and copper. It's SO cool. All that shiny metal, the airships and pneumatic trains, and most of all: the steam engines. Any future that requires boilers to be fed mountains of coal is a future I can be comfortable in. Steampunk is warm, and regal. It reveres a convergence of technology and Victorian opulence that promises both comfort and convenience, and it's a plus that this basic technology will be easier to start with as you rebuild society.

This experiment will help you learn about steam power in a neat way—through steampunk!

EXPERIMENT	STEAM POWER, STEAMPUNK STYLE
CONCEPT	Build your own steam engine, with a little steampunk flair, and test some properties of steam power.
COST	$—$ $
DIFFICULTY	⚙ ⚙ — ⚙ ⚙ ⚙
DURATION	☼ — ☼ ☼
DEMONSTRATION OR EXPERIMENT	Mostly a demonstration of the principles of steam power, but could involve experimenting with different designs.
TOOLS & MATERIALS	• 2'1/8" copper tubing (available at Amazon—extra is good for practicing bending the tube) • 1 tea light candle • 1 aluminum soda can or beer can • Durable scissors • Hole punch (optional) • Aluminum tape (optional) • Long-stemmed matches or a candle lighter • Basin or bucket of water (6-inch + deep)

The steam engine has been around in one form or another for thousands of years. The basic concept behind steam engines—which involves heating up water until it turns to vapor and expands, creating a pressure that when released can be transformed into mechanical energy—has been around forever and is used to create different kinds of energy. Heck, even nuclear power plants use the nuclear reaction to heat water and create steam to spin turbines and generate electricity. It's a ubiquitous way to get energy out of heat.

As projects involving fire go, this one is pretty tame, but still, please use caution. Older kids should be fine, but always supervised (and even the most careful geek parents can get singed by steam as well).

BUILDING THE STEAM ENGINE

STEP 1: Cut a soda can horizontally about $\frac{1}{3}$ of the way up. Either fold down the edges, or if you're feeling particularly worried about sharp edges, use some aluminum tape and wear gloves the whole time. Keep in mind the can will get pretty hot during the experiment, so never pick it up with bare fingers.

STEP 2: Melt some candle wax in the very center of the bottom of the can. While the wax is still warm, press a tea light down on top of it, anchoring it in place. Make sure it's centered as much as possible, as balance is important for the engine.

STEP 3: Measure out about 1 foot of copper tubing and bend it in the middle into two circles, leaving the extra on either side (try to keep them even as much as possible). It should look a little like those hand-exercisers you can buy to strengthen your grip, which use a v-shaped spring.

You will have to be very careful bending the copper tubing because it is very soft and prone to crimping and breaking. Try rolling it around a hard surface like a wine bottle neck to get the shape right, and warming it a bit in your hands before starting. It's best if you buy enough tubing to practice a few times. The more quickly you move, the less likely it is to break off. Trim excess if needed using needle-nose pliers, or simply bend it off (make sure not to crimp too much).

STEP 4: Punch a hole in either side of your can (you may be able to actually use a paper hole-punch to do this, as the aluminum is not very thick) and slip each end of the copper tubing through one of the holes, from inside to out, and pointing each of them downward so the coil of tube will sit above the candle flame when it is lit. Make sure there's enough copper pipe poking through on each side to hang below the bottom of the can by at least an inch.

STEP 5: Bend the extra tubing to a 90-degree angle tangent to the can, but in opposite directions (picture in your mind that jets of water are coming out of these tubes; if you were hanging the hole construct by a string tied through the coil, the jets would cause it to spin—which it's actually going to do!). The bent jets need to be at the bottom of the can or lower, to ensure that it is completely submerged when the can is floated in water.

STEP 6: Water time! Fill up your basin or bowl. And prime your pump. What does this mean? Well, you need to get water into the copper tube. There are a few ways to try this. You could hold one end under a faucet until water comes out the other, then quickly cover both ends and float the can to keep the water in. Or you could use an eye-dropper. The important part is to get the water in the coil, and the can into the basin with the jet ends submerged so the priming water doesn't escape.

STEP 7: Light the candle (carefully, as this can be tricky, what with it sitting in water and all) using long matches or a long candle lighter. Make sure the flame is pretty much directly under the copper coil.

STEP 8: Wait and watch. You'll notice a few bubbles at first, which is a good sign. As the copper heats up, the water inside will get warmer and warmer. As it turns into steam it will expand violently, eventually popping and sending the little vessel in circles.

That's it for the basic demonstration of steam power, but it's obvious you can take this project much further if you like. Instead of a can, you could build a boat, perhaps by bisecting a can and shaping it. Then set your coil and jets so they both point backward, and voilà! You have a steam jet-boat!

In terms of experimentation, you can try various designs with one, two, or more coils, and see how the different capacities increase rotation speed or duration of spin. You could (if you're ambitious) even measure the mechanical energy being created, if you know the weight of the can, candle, and coil, and you measure the number of times it rotates in a full circle every minute. A little math and a few assumptions later, and you'll get energy!

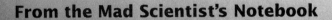

From the Mad Scientist's Notebook

If you really do want to calculate the energy, you'll need the following equations:

$$E_r = I\,\omega^2$$

$$I = mr^2$$

E = Rotational Energy

I = Moment of Inertia

ω = Angular Speed

m = Mass of the can

r = radius of the can

Just make sure you get your units all right!
Check out http://en.wikipedia.org/wiki/Rotational_energy to help.

Just never forget the steampunk look. All your robots must be built using brass rivets, and produce billowing clouds of black smoke from stacks coming out of their heads. Even if they're not actually burning coal. It's all about the look.

Mastering Alchemy

Idea by Kathy Ceceri

ere's the scenario: double apocalypse! A giant rogue asteroid is hurtling toward Earth, and for the first time in recorded history all of the countries in the world have joined together to launch a huge barrage of nuclear weapons out into space to deflect/destroy it.

Except it doesn't work. Damned metric conversions!

And so, now the 721 major pieces of the broken-apart-but-not-destroyed-and-now-radioactive-asteroid have come through the atmosphere and wiped out 95 percent of humanity. Plus they have melted all the ice on both poles, raising sea levels so high that every major population center, save a few, is under water (La Paz is doing okay).

It's an interesting time to be a mad scientist.

Of course, things are back to a barter economy. Certain metals are quite scarce (everything's under water), and what's left fetches a high price as people need metal to repair the electronics that have survived. Like copper. It was a lucky thing you'd squirreled away your penny collection in your secret laboratory and fortress hidden in a mountain for just such an occasion.

You've got a lot of pennies, but not enough to fund your rise to power as the Emperor of New North America, so you're going to have to stretch your funds a little. Luckily you know science, and have access to three very important things: salt, vinegar, and a nearly unlimited supply of paper clips (pilfering your local Office Depot just after the asteroid impact was a stroke of genius).

With these supplies, you're going to be able to figuratively print yourself money! Well, really, you're going to electroplate paperclips so people think they're made of copper. But no one will figure that out until it's too late!

Muahahahahaha!

EXPERIMENT	MASTERING ALCHEMY
CONCEPT	Using some simple household materials, we're going to dissolve copper pennies and electroplate them onto other metals.
COST	$
DIFFICULTY	⚙
DURATION	☀ ☀
DEMONSTRATION OR EXPERIMENT	Demonstration of science
TOOLS & MATERIALS	• ¼ cup vinegar • 1 teaspoon salt • 20 pennies • Iron nails (steel paperclips work, too) • Disposable plastic cup or bowl • Plastic spoon

Back in the days before we even understood atoms and molecular structure, people called alchemists searched for a way to turn one element (lead, for example) into another (gold, for example) because it would be a huge increase in the metal's value. But while it was born out of magic and superstition, alchemy became the basis of the science of chemistry. Many of the tools and techniques first developed by alchemists in the Middle Ages are still used in chemistry labs today.

You can't turn one element into another, at least not without a nuclear reaction and/or a particle accelerator (let's welcome the newly created elements 114 and 116 to the periodic table, everyone!—they were created in laboratories by smashing smaller elements into bigger ones), but through the "magic" of chemistry you can make iron (Fe) look like copper (Cu). The trick is to coat the iron with a thin layer of copper, a process called copper plating.

STEP 1: To start, we're going to mix up a strong acid solution. Okay, not *that* strong. It won't be alien-blood-eat-through-the-floor acid. Just something we can dissolve some pennies with!

Pour $\frac{1}{4}$ cup vinegar into the plastic bowl. Stir in 1 teaspoon of salt. The vinegar is a solution of acetic acid (CH_3COOH) and water (H_2O), useful as a weak solvent or an ingredient. Salt (NaCl) is made up of sodium and chlorine. When you mix them, they combine to form a slightly stronger solvent called hydrochloric acid (HCl).

STEP 2: Use the acid to dissolve copper from the pennies! Put 20 dull pennies in the bowl with the acid. Let them soak for about 5 minutes, then remove them and let dry on a paper towel.

STEP 3: Now we're going to turn some iron nails or steel paperclips into "apparently" copper nails or paperclips, using the acid solution.

Once the pennies have been removed from the bowl of acid, add the nails or paperclips (leave one out for comparison so you can see the difference when you're done). You can also stand or prop one item up so it is half in and half out of the solution.

Look closely and you will see bubbles forming on the surface of your items. This is a sign that a chemical reaction is taking place. The bubbles are hydrogen gas being released by the chemical change.

Here's the hard part: You'll have to leave them for a while, like a couple of hours or more. But after that time, the nails or paperclips should turn a bright coppery orange. You did it! You're an alchemist!

Well, not really. You're a copper-plater instead. What really happened is that, just like the copper from the pennies, some of the oxidized metal on the surface of the nails or paperclips dissolved in the acid solution, releasing positively charged iron ions (Fe^{+2}). Meanwhile, the extra electrons left behind gave the surface of the nails a negative charge. Because copper is more "noble" than iron (a chemical term meaning that copper goes into a dissolved state less easily than iron), the copper ions were more strongly attracted to the iron nail than the iron ions.

What Else Can I Do?

For an additional chemical reaction, repeat steps 1 and 2, but try rinsing some of the pennies in water before drying. The rinsed pennies will remain shiny clean. Un-rinsed pennies will start to turn green as they dry. This greenish coloring, like the copper of the Statue of Liberty, is called *verdigris*. It is a layer of copper carbonate ($CuCO_3$) formed by a reaction with the acid.

But what happened to the dull coating on the dull pennies? Did we go through that just to wash them? Nope! That dull coating was a layer of copper oxide (CuO), formed when the copper on the surface of the penny reacted with the oxygen from the air (it's the copper version of rusting). The HCl dissolved the CuO layer to release copper ions (Cu^{+2}).

Ions are atoms that contain more or less than the normal number of electrons. The copper ions in the vinegar solution are positively charged because they left two of their (negatively charged) electrons behind when they separated from the rest of the copper penny.

Here is the chemical equation that shows this reaction: $3(CH_3COOH) + H_2O + NaCl + CuO = 3(CH_3COO^-) + 3H^+ + 2OH^- + Na^+ + Cl^- + Cu^{+2}$

True copper plating is done with much more powerful chemicals, and produces a much smoother, sturdier coating. What you've created can probably be scratched off, so it won't hold up under serious scrutiny. But it should be enough to make a trade with some unwitting survivors of the apocalypse for other valuable materials and equipment!

Beyond that, knowing how to copper plate is cool. If the future is steampunk and retro futuristic as we hope, being able to add a shiny copper coating to all your iron and steel snaps and zippers will make you the best decked-out mad scientist around.

The Science of Siege Warfare

hat's old is new again, and that's never truer than when you're trying to outfit your army for your post-apocalyptic takeover of the world. When most of the heavy weapons have been destroyed by the flood waters or taken by the alien scavengers, it's time to go the DIY route.

So, we turn to physics and we turn to history. The trusty trebuchet is the perfect example of the laws of motion cunningly applied to a machine for war. It's also a work of art the way it performs its little pirouette to launch. It's magical. It's the iSiege weapon!

EXPERIMENT	THE SCIENCE OF SIEGE WARFARE
CONCEPT	Build your own trebuchet and test it to find the combination of factors that optimize its capabilities.
COST	$—$ $ $
DIFFICULTY	✿✿—✿✿✿
DURATION	☼ ☼ ☼
DEMONSTRATION OR EXPERIMENT	Experiment to study the physics behind a classic piece of siege warfare machinery.
TOOLS & MATERIALS	• Measuring tape • Chalk • Video camera (optional) • Tripod (optional) • Roll of butcher paper • Marker pen • Trebuchet You can purchase a trebuchet kit, or build your own with the following parts: • LEGO bricks • LEGO Mindstorms NXT or Technics building pieces • String • Towel or other fabric • 100 ¼-by-1½-inch fender washers • Key ring.

This project is based on the one my older son Eli did for his seventh-grade science fair. He built his own trebuchet from a kit, then chose three differing shapes of ordinance and launched each one 10 times, recording the distances thrown. Using that data, he validated his hypothesis about which shape would fly the farthest.

There are many other possible experiments to be done with a trebuchet. But first, you have to make one. Of course, you can purchase them pre-built, but really the fun is in the making. There are a number of kits available online, from places like trebuchet.com. I also have a fondness for the snap-together kit produced by a pair of industrious young men, called the trebuchette (go to kickstarter.com and search for trebuchette). In another of my books, *The Geek Dad's Guide to Weekend Fun*, I had instructions for how to build your own trebuchet from LEGO parts. I include those instructions here for your use. However, the instructions do include some <SPOILERS> that give away bits of information you may want to suss out through experimentation later on.

BUILDING A LEGO TREBUCHET

STEP 1: Building the Lever Arm

The lever arm needs to be a pretty rigid construct. You can use the beam pieces that come with the NXT set, along with the joining brackets, to achieve this. Regular LEGO bricks won't work, because the torsion that will be applied by the counterweight will overcome the way bricks lock together. What you need are pieces that will lock rigidly.

The length of your lever should be about $\frac{1}{4}$ for the counterweight and $\frac{3}{4}$ for the throwing arm. There must be a way to attach the lever arm to your superstructure at this $\frac{1}{4}$ and $\frac{3}{4}$ fulcrum point that will allow it to swing freely, like an offset seesaw.

For the counterweight, I put two simple crosspieces between the beams and threaded enough stamped washers to fill all the way across (at least 20 for each beam, for a total of 40). The whole idea is to get a pretty big weight on the short side of the fulcrum, so instead of washers, if you have something like lead fishing weights, which will be even heavier for the space they'll fill, that's a good upgrade. Actual numbers will depend upon the scale of the trebuchet you build, but the ratio of your counterweight to the weight of your projectiles should be around 10 to 1, or more.

STEP 2: Building the Superstructure

Here's where the traditional LEGO bricks come in most handy. All the superstructure has to do is support the weight of the lever arm and counterweights at rest, and be able to deal with the rotational forces when it's in motion.

Use a variety of 2 × 2, 2 × 3, 2 × 4, 2 × 8, and even 2 × 10 bricks to build the structure walls for supporting each side of your lever. Make each structure wall 2 studs wide by 12 studs long so you can maximize the combinations of bricks available. The most important thing is to stagger seams. If seams line up on multiple levels, they weaken the overall structure. If you can, build the walls on a LEGO base plate. Separate them by the width of your lever arm construct.

The height of the support walls depends on the length from the axis point of your lever arm to the outside edge of your counterweight. Since the counterweight is going to swing through an arc, the axis point has to be high enough so the low point of the swing will still keep the counterweight off the ground.

To attach the lever arm to the structure walls in this design, use a Technic-style brick with holes in it, through which a crossbar can run. This allows the lever arm to rotate freely.

An additionally helpful design feature for the support walls is a crossbeam near the stop, to help keep the walls vertically true and to act as a stop for the lever arm when it finishes its rotation to launch of the projectile.

STEP 3: Building the Sling

The sling is the heart of the trebuchet. It's the idea originally created to extend the throwing ability of the human arm. On a trebuchet, the lever replaces the arm, but the sling remains the same.

There are three components to the sling: the pouch, where the projectile rests; the cords; and the thimble, which is the release that allows the trebuchet to actually launch the projectile.

For the pouch, I just cut a rectangle about 2 inches by 4 inches out of a shop towel, and then trimmed the corners off. At the ends, I punched holes so I could tie them off to the cords.

There are two cords. The cords should be made of string for this build. Each one should be about the length of the long section of the lever from the axis to the end. If you ever decide to build on a larger scale, you could use lightweight rope instead.

One end of each cord ties off to an end of the pouch. One cord then ties off to the end of the long part of the lever. The other cord's end gets tied off to the key ring, which will act as the catch-hook that keeps the whole launch package swinging around until the top of the arc. Both cords should still be of roughly equal length, so that when a projectile is loaded in the pouch, it will hang from the lever arm evenly.

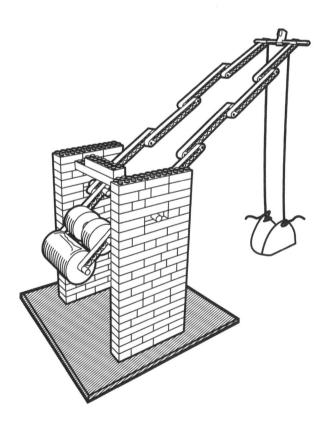

STEP 4: Fetchez la Vache

Now it's time to have some fun launching projectiles and seeing how far they can go. Remember, you may need to do a bit of fine-tuning to get everything running smoothly—your projectiles should be a very small percentage of your counterweight. I used the small colored balls that come with the Mindstorms NXT set, but you could easily use standard 2 × 2 LEGO bricks.

To launch, hook the key ring end of your sling over the catch at the top of the lever arm. Now load your projectile into the sling, then pull downward and inward under the fulcrum so you are pivoting the counterweight upward on the opposite end. When you let go, the counterweight will pull down, pivoting the arm, which draws the sling out, around, and over the top.

When the sling flies over the top, the key ring slips off the catch, allowing the projectile to escape and launch into the air. Your sling has to allow the key ring end to slip off the catch at the optimal part of the lever rotation so the projectile launches at a roughly 45-degree angle to the ground. Tweaking each part of your build can maximize the effects. But it's the perfect way to learn from trial and error. </SPOILERS>

DEVELOPING YOUR EXPERIMENT

Now that you have a trebuchet, what do you do with it? What kind of experiments can you conduct? Well, first think about how it works.

The trebuchet works on a couple of principles. There's the lever arm that demonstrates Archimedes' groundbreaking work, but in reverse. The ancient Greek mathematician explained how one could use a lever to lift something very heavy using a much lower amount of force than directly picking it up would entail. Ironically, a trebuchet uses a much heavier object and a lever to pick up a lighter object (and throw it a long ways). It's all the same forces at work, just a different application.

The second principle at play is that the sling extends the length of the throwing arm, delivering a greater amount of rotational force for a more significant launch distance. Just like a person can use a sling to throw a stone much farther than they could using only their arm, tacking a sling onto the arm of the trebuchet makes it a very potent device for launching projectiles.

What's My Hypothesis?

Knowing the two fundamental principles behind how a trebuchet works, we can consider the variables to study:

▶ The ratio of the counterweight to the projectile.
▶ The ratio of the length of the counterweight arm to the throwing arm.
▶ The length of the sling with respect to the throwing arm.
▶ The angle of release of the sling.
▶ The density of the projectile.
▶ The shape of the projectile.

Maybe you can think of a few more? In any case, choose something that can be varied, and form a hypothesis around it. For example, as I mentioned above, my older son Eli chose to test different shapes of projectile to see which would travel farthest. He controlled any other variables by making sure each projectile was made out of the same material, and weighed the same amount. He hypothesized that the greater the number of sides on the projectile, the farther it would travel, both because of reduced air resistance, and travel after landing. I won't tell you his results, but I can say he got an *A* on the project!

SETTING UP THE EXPERIMENT

The challenge of experimenting with a trebuchet is recording the results, especially if you're working by yourself. That's why I'm going to suggest a couple of different methodologies for performing the experiments that can make it easier.

Your first choice, and one which is faster to do, will require two people, and a place like a schoolyard or wide sidewalk where you can work, and where no one will mind if you mark up the paved surface with chalk. Working with a partner, make a few test launches to make sure the equipment is working, and to give your partner (who will act as your "spotter") a good idea of the landing zone. You should also decide whether you're measuring the distance to where the projectile first lands, or where it comes to rest—each figure could be important, depending upon your hypothesis!

With your projectiles ready, and your spotter in position, you're ready to start making launches. But you need a plan to make sure data gathering will be efficient and accurate.

Say you have three different projectiles to test launch. Assume you'll launch each one 10 times (minimum) to establish a good data set. You can identify each launch by a projectile number and a launch number. Using the first projectile, the first launch would be P1L1, the second P1L2, and so on up to P1L10. Then start with projectile 2 at P2L1 and carry on. When you're done, you'll have made 30 launches, from P1L1 through P3L10.

Before each launch, the trebuchet operator (you?) should call out the launch ID, like "P1L3!" The spotter should repeat it back. The operator counts down and launches. The spotter tracks the shot, and when it lands (either first bounce or final resting place, as we discussed on the previous page), marks an "X" on the ground with the chalk, and writes the launch ID right next to the mark. When done, there should be 30 marks on the ground with launch IDs next to them.

Now you can record your data. If you were thinking ahead, you might have set yourself up a table like this:

LAUNCH #	PROJECTILE 1	PROJECTILE 2	PROJECTILE 3
1			
2			
3			
4			
5			
6			
7			
8			
9			
10			
Total Average			

With the help of your spotter, use the measuring tape to record the launch distance for each shot onto the table, then take it home to do your analysis.

Alternately, if you're working by yourself, you need technology to help you a little. Roll the butcher paper out to be your landing field. Use the measuring tape to draw distance lines all the way down the paper with clear, heavy lines at every foot, and strong dashes for each inch. Mark the distance at every foot line in large numbers.

Set up your video camera on the tripod at the end of the landing area (make sure it is out of range from being hit by flying projectiles). The camera should be pointing slightly downward so the trebuchet and the entire landing area (especially the lines and numbers) are visible and readable. Start recording.

Prior to each launch, you should call out the launch ID loudly enough for the camera to pick up the sound. Go through all your launches. Then check the camera to make sure the entire session was recorded.

Now, instead of measuring everything out in the field, you can take the video home and review it on your computer or television, and enter each shot onto your table by watching where they land in the recording.

ANALYSIS AND CONCLUSIONS

The data for this kind of experiment is thankfully pretty easy to deal with. You basically only need to average a given set of launches to develop a result with which to challenge your hypothesis. Depending upon the level of rigor you'd like to apply, you could consider dropping high and low values before averaging, or including a look at the range of your values to make sure you account for any outlying results.

The best part is, of course, getting to play with a siege weapon, and learn a little lesson about history and technology. The ingenuity of the trebuchet is enough to impress, even today, and it's always good to know how to build and use a device to defend yourself that needs no electricity or explosives. Because who knows what we'll have handy when the alien scavengers come calling?

Post-Apocalypse Particle Detector

Idea by Kathy Ceceri

For the mad scientist working diligently to put human society back together after the apocalypse (they haven't found out you started the whole thing yet, have they? No? Whew!), there are a few important issues to take care of from the outset. Food and water are vitally important for your new "minions" (or shall we say "survivors depending upon you for their very lives?" Yeah, better PR that way), but perhaps even more important is making sure that food and water, and the ground they're growing it in, and the blocks they're building your new palace out of, are all safe. And by safe, we mean free of the radiation generated by those alien death rays, or the accidentally triggered doomsday bombs.

And let's assume the death rays or bombs took out all your radiation-detecting equipment as well. You're just as in the dark (figuratively, and sometimes literally) as all the others. Well, maybe not quite as much, because you are, after all, a master of science! And you know how to scrape together a particle detector out of just a few common tools and materials.

EXPERIMENT	POST-APOCALYPSE PARTICLE DETECTOR
CONCEPT	You can detect subatomic particles from space, the environment, or radioactive material by watching for ion trails in clouds of alcohol vapor.
COST	$ $ (or less, depending on radiation source)
DIFFICULTY	⚙ ⚙
DURATION	☼ ☼
DEMONSTRATION OR EXPERIMENT	Demonstration; experiment can test different designs, possible sources of radiation, etc.
TOOLS & MATERIALS	• Heavy, wide-mouth glass jar with lid. Can also use a Pyrex glass dish with rubber top. • Black peel-and-stick felt sheet (available in craft stores) • Black construction paper • Tape • Isopropyl alcohol, as pure as possible (you can get 91% pure in drugstores) • Styrofoam cooler with loose lid • Heavy gloves and/or tongs • Dry ice (sold in some food stores and welding supply stores, $1–$2 per pound, usually 10-pound minimum; easier to find around Halloween) • Radiation source such as uranium marble ($10/3 at unitednuclear.com), radium watch hand, cosmic rays, smoke detector, potassium salt, banana • Washcloth • Plastic wrap • Microwave oven • Bright flashlight

What we are going to build in this project is a particle detector known as a *cloud chamber*, which makes the particles shooting off a radioactive material visible. Scottish-born physicist C. T. R. Wilson invented the cloud chamber in 1911 (it's often also referred to as the Wilson Chamber), in order to study optical phenomena in clouds, when he discovered what else his device could do. He was awarded the Nobel Prize for the discoveries he made with it in 1927.

What's the Science Here?

Radioactive atoms are so unstable that they decay, or break apart into other elements. Part of that breaking apart results in the release of subatomic particles. This release is referred to as radiation; literally the radiating off of particles. Even when no known radioactive material is present, radiation from outer space and the environment is all around us.

Indeed, you may have heard trivia about how people who fly across country often receive as much or more radiation than people receiving x-ray imaging (true fact!). This is because although much of the radiation coming from space (aka cosmic radiation) is stopped by our atmosphere, the higher up in the air you are, the less atmosphere there is between you and space to shield you from cosmic radiation. That's one reason why the space shuttle, and the space suits that astronauts wear, are so well shielded. When they're in space, there's nothing else to shield them.

The cloud chamber is filled with a fog of alcohol vapor (not for drinking or inhaling!). The fog is produced when you create a difference in temperature between the top of the particle chamber and the bottom (a temperature gradient). Heat causes the alcohol to evaporate (change state from liquid to gas), and extreme cold causes it to quickly condense back into a liquid, in the form of little drops of fog.

The radioactive particles in the chamber (from our various sources) are electrically charged. They ionize the air molecules in the chamber as they pass through, and the alcohol condenses around them. This leaves behind an *ion trail* that can be seen in the cloud and that looks like moving streaks.

Our version of the cloud chamber is going to be pretty simple. It will involve building a testing chamber and creating the alcohol fog that will let us see the ion trails created when the ions pass through. You'll want to do your experiments in a room that can be made as dark inside as possible. A basement or bathroom without windows would be perfect, or another room where heavy curtains can cover the windows.

STEP 1: First, we need to accessorize our cloud chamber so we'll be able to see the ion trails. Your choice of jar or dish will be important for visibility. Obviously, the larger the chamber, the better. Tall and wide is nice.

Cover the outside of the jar or dish making up your chamber with black construction paper, with two "windows" cut out: leave one window big enough to see in, half as big as the side of the vessel that you see from straight on, and centered (so, if you have a jar that's 10 inches tall, and 10 inches in diameter, make the window 5 inches by 5 inches). Add a second, smaller window (say $\frac{1}{3}$ of the first window) a quarter of the way around from the big window to shine a light in from the side. Also, cut a piece of construction paper to fit inside the bottom of the jar and place inside. You've blacked out your chamber, so that with the lid on, the only way for light to get in or out will be from the two windows.

STEP 2: Cut a piece of black peel-and-stick felt to fit the inside of the lid of your chamber, and attach. The felt will be soaked in alcohol when we do the experiment, and provide the fog in which we'll see the ion trails.

STEP 3: We need to prepare the dry ice prior to the start of the experiment. Put the lid of the Styrofoam cooler you're storing the dry ice in upside down on your work surface. Place a piece of the dry ice into the upside down lid. If you have a large chunk of dry ice, break off a slab-like piece that will fit inside the lid, and be as level as possible. If you have dry ice pellets (available in some places), fill the inside of the lid with a nice even layer, as deep as possible. We're going to place the testing chamber on top of the dry ice to chill the air inside, so we need a flat, stable surface. You can place a thin metal

Important Safety Notes

As we've discussed before, safety is the first order of business with any project. This particular project contains several materials that require careful handling to be safe:

- ► Dry ice, which is solid carbon dioxide, has a temperature of –109 degrees Fahrenheit. All experimenters must use appropriate gloves or tongs when handling, and wear safety goggles or safety eyewear when breaking up large pieces. The dry ice should be stored in a Styrofoam cooler with a loose top. Never seal the lid tightly. We'd rather there be a vent, since pressure can increase as the dry ice warms up and causes it to subsume into gaseous CO_2.
- ► Isopropyl alcohol is flammable and gives off strong fumes. Use gloves to avoid getting it on your skin, and use with ventilation. NEVER have an open flame near where you are working with the alcohol!
- ► Uranium marbles (if you use them) should be kept in a sealable plastic bag away from small children and pets. They are safe to handle, but you should wash your hands after handling them (better: wear gloves!), and avoid getting chips or dust in your mouth or eyes.

cookie sheet directly on top of the dry ice to make a more stable surface, if needed.

STEP 4: Now we're ready to prep the chamber and start the experiment. Soak the felt in the lid of your chamber thoroughly with the alcohol (you may want to work near a window, or outside, to avoid fumes). Place your radiation source inside the chamber and close the lid.

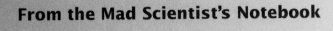

From the Mad Scientist's Notebook

While radioactivity and radiation have quite a lot of negative press as dangerous, bad things to be avoided, the truth is (as we've mentioned previously) there is radiation around us all the time, every day, and there are many sources we can use for this experiment:

- Uranium marbles are cheap (3 for $10) and relatively safe to handle (again, use gloves, or make sure to wash after handling). You can get them from http://www.unitednuclear.com and on eBay (though more established science suppliers will likely provide better-quality materials). You can also buy a more potent "radiation source" needle from United Nuclear, but they are much more expensive, and perhaps not something you want just lying around the house after you're done with the testing.
- Salt substitute, known by its chemical name as potassium chloride, contains measurable amounts of radioactive potassium-40.
- Bananas. Yes, bananas! They also have the above-mentioned radioactive potassium-40. You may have eaten a source of radiation today!
- Some smoke detectors contain radioactive Americium (which is not, mind you, the secret material in Captain America's shield). If your test chamber is big enough, put the detector inside for the test. If not, set it next to the chamber and run the test. I can't, however, recommend removing the Americium from the detector. Hacking radioactive materials is not the smartest thing in the world.
- If you have no radiation source, you can still watch for cosmic gamma radiation, but you should probably try a larger cloud chamber to increase your chances of seeing rays.

STEP 5: Get the washcloth wet, fold it up a few times until it is about the size of the jar lid, and wrap it in plastic wrap so it doesn't leak (alternately, put it in a plastic sandwich bag, but DO NOT SEAL IT CLOSED). Place your wrapped wet washcloth in the microwave for 15–30 seconds, or until it is about body temperature (we don't want it too hot, or the fog won't form effectively).

STEP 6: Put the testing chamber on the dry ice, as level as possible with the observation window facing you. Put the warm washcloth on top of the lid. Position the flashlight or other light source so that it shines in through the side window.

STEP 7: Turn out the rest of the lights. Wait a few minutes until a rainlike mist starts to form inside the testing chamber. You will begin to see lines streaking off from the radiation source or in random directions. If you're having trouble seeing any trails, try moving the light source around.

Different radioactive particles create different-shaped trails. Uranium produces alpha particles (basically rogue Helium nuclei), which leave short, thick trails in the fog. Other radiation sources produce beta (high-energy electronics or positrons) particles or gamma radiation (very short wavelength, high-frequency). Cosmic rays may produce crooked trails as particles collide with other atoms.

If you want to increase your chances of seeing cosmic rays, or test larger samples to get more spectacular results, try building a larger chamber. In 2007, Scottish high school student Holly Batchelor won the Intel International Science and Engineering Fair's First Award for the radiation detector she built out of a plastic aquarium.

Once you've familiarized yourself with how the different forms of radiation look in your cloud chamber, you'll be able to go out and identify radiation on random samples so you can know what's safe and what's not. And that knowledge may be the key to your success as a mad scientist, and ruler over all the survivors of that terrible, terrible tragedy that no one (except maybe you) could have foreseen!

Mapping Your Ecosphere

Idea by Kathy Ceceri

I n some far-flung post-apocalyptic future, we may all be living in glass-domed ecospheres, surviving via science while the earth repairs itself from the ravages of ecological collapse or wanton destruction outside. Each of us will have to understand the ecosystem around us, learn how to encourage the maximum output from our gardens, and make sure invasive species (like, say, zombie squirrels) don't sneak in and destroy our ability to sustain ourselves. What we can do is perform what's called a *biodiversity audit* on our ecosphere now, before the aliens/zombie squirrels/robot overlords come and mess everything up.

How many different species do you think live in your backyard or neighborhood? How many of them can you identify? According to the BBC, Charles Darwin conducted the first biodiversity audit in history in June 1855, when he began a study of the local flowers in the meadows around his home in Kent, England. In that case, he merely walked around his neighborhood and catalogued each plant he could find. Seems simple now, but it was the start of something important.

Today, the technology we have available allows us to make a far more detailed and informative audit. We can now map our local ecosphere in great detail, using the available online maps, spreadsheets, and the research materials in books and on the Internet.

EXPERIMENT	MAPPING YOUR ECOSPHERE
CONCEPT	A biodiversity audit helps you learn to identify different species, where they live, and how they interact with other species and their environment.
COST	$
DIFFICULTY	🔧
DURATION	☀—☀ ☀ ☀ ☀
DEMONSTRATION OR EXPERIMENT:	Either. As a demonstration, it involves observation and presentation of data. Can also use it to test a hypothesis about living things in the environment, such as where the greatest number or different types will be found, or how changes are affecting them.
TOOLS & MATERIALS	• Notebook • Pen or pencil • Map • Wildlife field guide(s) • Internet access Optional: • Camera • Binoculars • Sample bottles or bags for microscopic life forms • Microscope • Digital drawing program • Web site, blog, or online photo album

STEP 1: Start with Cartography

First, you need to decide which area you're going to study. This can be as simple as your backyard, or a nearby park or field, depending upon where you live. You want to consider your ability to access every square foot of the space, whether you can easily delineate the space with paths or fences on each side (so you know where the ecosphere starts and ends), and if there is some vantage from which you can overlook the area so you can get a good overview.

But the most important thing is to choose an area with signifi-

What's My Hypothesis?

If you want to make this a classic science fair project, the best choice for a hypothesis is pretty easy: Just ask yourself (formally) how many different species you think live in your yard or neighborhood. Come up with an estimate, and then perform the audit with detailed mapping as your experimentation. A somewhat more detailed hypothesis might involve suggesting which species you expect to find, and in what quantity. Remember: Maps, pictures, and matrices of data make for great science fair visuals!

cant biodiversity; meaning there's lots of nature there, like birds, insects, wildflowers, weeds, vegetable gardens, fish ponds, mushrooms and slime mold. Pick an area (backyard, street, park, etc.) that has as many different kinds of environment as possible. A wild meadow will have more different species than a manicured lawn.

You could even leave patches of your backyard unmowed for periods of time in order to create little minimeadows to study. Depending upon where you live, you may be amazed and entranced by who moves in (just as long as they're not zombie squirrels).

Once you choose the area to be surveyed, make a map showing landmarks such as buildings, sidewalks, trees, fences, gardens, streams, and ponds. It's up to you to decide the level of detail you want. You could make a hand-drawn map, with only the obvious details. Or you could copy an online satellite image, and then use a graphic program to overlay labels and lines of demarcation.

On your map, divide your ecosphere into sections and give each section a label, perhaps something as simple as a letter for each area. Areas/items that may be designated as individual sections include lawns, flowerbeds, trees or bushes, pools or ponds, decks, hills, patios, and more. Try to segregate contiguous zones; meaning: If a type of area is an expanse of one thing, make it one labeled section (like a lawn would be one area, a pond another, a bushy area the third).

STEP 2: Recording Species

You're almost ready to head out and record some data (woo-hoo!). But thinking ahead and planning always make for a smoother experiment. What will really help you in the field is to make yourself a matrix to fill out with the data you're collecting.

You already know how your zones are identified (with letters), so for this part, let's assume you'll assign a number to each species you find in a given zone. You may find the same species at a different point in your audit of subsequent zones, and it doesn't matter if their number from one zone doesn't correspond with their number in another zone. If Species 3 in Zone B is the same as Species 7 in Zone K, it won't be a big deal (unless they're zombie squirrels). You'll note instances like these when you analyze the data later.

So, with all this in mind, build your matrices. Here's an example:

ZONE A SPECIES LIST 1						
	DATE	TIME	PLACE	TYPE	NAME	DESCRIPTION
1						
2						
3						
4						
5						

The matrix identifies which zone is being recorded, the number of the species identified, the date of discovery, the time, more precise location, the type of creature (in broad terms: bird, flower, varmint), the name of the species (if you know it right away), and a description. You could create this matrix in a spreadsheet program on your home computer and print out one for each zone, or you could lay it out with pencil and ruler on some graph paper, and then photocopy a bunch (leaving a space on the Zone ID for you to fill in for each one).

Now you're ready! Slap those blank matrices on a clipboard, stuff your pocket with #2 pencils, grab a camera, magnifying glass, and/or binoculars (if handy), and head to the starting line. Or, in this case, the starting Zone.

Think of this part of the experiment as CSI: Your Backyard. The key here is to be deliberate, meaning you should look around slowly, carefully, and with an intense attention to detail. Especially in planted or "wild" areas, you're going to want to look very closely, and especially underneath plants and rocks or up in the top branches of a tree (hence the binoculars), to find every last living being you can. Indeed, it's a bit like a treasure hunt, except that the treasure is everywhere!

Of course the tricky part is knowing what you're seeing. Once we get past things like *sparrow* and *garden snail* you may start running into some stumpers. Yes, those are ants, but which kind? How many of those seven kinds of ice plants in the "low-maintenance" section of your garden can you actually name? No, you're not going to have all the answers at your fingertips.

So, you have a few options. First, see if you can get your hands on the wildlife and planting guides for your area from the library or bookstore. You can carry that with you and look up most anything you might run into. If you don't

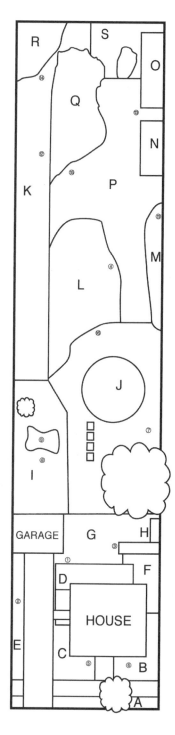

have a guide, write down as many details as you can about the specimen. Maybe even draw a sketch. Get as much information as you can for a later attempt to peg it via Internet search.

If you have a camera, go ahead and take a photo of every species as well. If your camera has it, use a macro mode so that the details of the specimen you're focusing on stand out. If you're taking pictures like this, do your best to download the pictures to your computer and put the "Area" and "Species" designations (letter/number) in the file name while you can still remember which is which (another reason to take notes on appearance, even if you're taking pictures).

You can even try to include microscopic species (good if you have a natural pond in your research area). Take a sample of water or soil (keep it in a sterilized container to prevent cross-contamination), and examine it under a microscope later. There are even microscopes that will hook up to your computer and display the image there, where you can save them.

If this is a one-shot for a science fair project or family learning, you can probably do the survey over a few hours, and then spend the bulk of your remaining time researching and identifying species and all the follow-up in the next few steps. However, this project can be bigger and more long-term than that. You could repeat the survey on subsequent days (good idea to track the weather conditions as well, which can have interesting correlations with the species present). You could make it a seasonal audit. Or to make it truly long-term, repeat it every year and track whatever macro changes are made to the research area as well (landscaping changes, large-scale weather impacts). There are a lot of possibilities for the data-gathering part of this project.

STEP 3: Identification

You need to identify any of the species catalogued in the field that you couldn't do while you were out there. If you didn't have any books or the Internet sites handy to do it at the time, that's your mission now. And remember, the resources at your disposal are probably pretty significant. Your local library likely has books about the ecology of your region. If you have a local parks department, they should be able to tell you a lot. There may even be ecological organizations in your area that know everything you need. Indeed, this part of the project is as much about searching out resources as the previous part was about finding the species themselves.

But even more than identifying the specimens, this is your chance to learn about them. As a matter of family coeducation, this is a way for you all to become closer to your home and environment. As a science fair project, this is where you get to demonstrate depth of learning that can make the difference between two colors of ribbon.

STEP 4: Crunching the Data

So, now that you have all this wonderful data, what are you going to do with it? Well, first you have to validate or invalidate your hypothesis, if you made one at the start of this project. How many species did you find? Is that more or less than what you guessed? Were the types of specimens what you expected, or were there some that totally surprised you?

Obviously, you'll need to tally everything up. Spreadsheet programs are really handy for this kind of thing. Once you can identify everything you observed, in what quantities, and where, you can start drawing conclusions. Identify the top 5 or 10 species in your study area and make a chart showing which zones they show up in. Can you glean any insights from this chart? Are there more flying insects close to a pond? Did you note birds near some trees?

And did you make any mistakes, like trying to search too large an area, or misidentifying any of the things you found? It's just as important to recognize those because as mentioned earlier, you can learn as much or more from your mistakes as your successes.

STEP 5: It's All About Presentation

If you are going to make this project a science fair presentation, you're going to need to present three parts: the hard data, pulled from your audit log and collected into a binder; the graphs and conclusions you've drawn from the data plotted neatly and arranged on a display board; and most importantly, the map and pictures. Print out an oversize version of the map to fit on one section of the board, so you can show off the zones and dots to identify species numbers in each zone and easily correlate the visual of the map to the data in the binder. And blow up your best pictures, arranging them around the display board with notes indicating which zones they are depicting. Make it all as visual as possible.

In the end, whether you do this as a science fair presentation, a solo project, or an experiment to get the whole family involved, auditing the ecosphere around you is great for learning about and connecting with your home and community. And how impressive is it to be able to identify the trees, plants, and animals in your neighborhood accurately and be able to talk about them with sound knowledge? It's pretty geek, and cool.

As long as you don't find a zombie squirrel.

More Info Available

Download and read this wonderful handbook published by a Welsh (as in Wales, UK) science society to encourage community biodiversity audits. While the scope of this book is a bit larger than a personal or family audit of the backyard, many of their guidelines and suggestions are entirely applicable.

http://www.scienceshopswales.org.uk/documents/Biodiversity%20handbook/handbook_web.pdf

FUN WITH FIRE AND ELECTRICITY

The Music of Fire
Building a Rubens' Tube

Idea by Anton Olsen

F ire. One of the original four elements, and understandably so. So
much power, so much destructive force.

But how a fire works—what colors it shows, how high or low
the flame is, or how hot it actually burns—can tell us things.

And in so doing, it can also be beautiful. Of all the experiments
in this book, building the Rubens' tube is the one that's as much
about creating a thing of beauty as it is about learning scientific
principle. Indeed, fire will light our way.

EXPERIMENT	THE MUSIC OF FIRE: BUILDING A RUBENS' TUBE
CONCEPT	Sound waves traveling through a gas create variations in pressure according to their wavelength. Creating a Rubens' tube demonstrates this, with fire.
COST	$ $—$ $ $
DIFFICULTY	⚙ ⚙ ⚙ — ⚙ ⚙ ⚙ ⚙
DURATION	☀ — ☀ ☀
DEMONSTRATION OR EXPERIMENT	This is a demonstration project, but a pretty visually awesome one.
TOOLS & MATERIALS	• 3-inch-diameter ventilation duct, 3 feet long • Duct tape • Glad Press'n Seal Wrap • Inexpensive propane torch and bottle • Amplifier and speaker (try to match speaker radius to duct radius) • Sound source (phone or computer) • 440 hz sine wave audio file. • Hacksaw • Drill (prefer drill press) • $1/16$-inch drill bit • Measuring tape, yardstick, or ruler • Metal snips (optional)

Understanding that sound is a function of pressure waves traveling through the air is a really interesting thing. It's all pretty invisible when it happens; you can't see the air moving from sound waves. Though you can do interesting things with other gases—like helium, which due to its lighter nature makes the same sound waves higher frequency (and most amusing). And the speed of sound varies depending upon the temperature and atmospheric pressure of the local environment.

So what can we do to visualize sound as waves in a gas? We can play with fire, and build a Rubens' tube.

A Little Science History

The device we're making in this experiment is a variation on an original experiment carried out by Heinrich Rubens in 1904, back when the concept of sound creating pressure waves in a gas was novel. Rubens and his predecessors John LeConte, Rudolph Koenig, and August Kundt had worked hard over the previous fifty years sorting out the wheres and whyfores, until Rubens decided to drill a bunch of holes in a tube, pipe some gas in, and light the gas so it made little flames all along the tube, then started making noises at one end. The end result was demonstrating how a modulated tone passing through the gas in the tube would set up a standing pressure wave that would show up in the flames. This is the way an oscilloscope (invented decades later) works.

And here, as I've included with all the fire/electricity/chemical experiments in this book, comes the safety talk.

This experiment uses a hacked-together apparatus to feed flammable gas into a tube, which you will then light, making a series of little candle flames. Things will get hot. Un-burned gas may leak. There is an element of danger in this experiment. Therefore, you must use caution and follow these guidelines:.

- ▶ Work in a well-ventilated space

- ▶ A parent, guardian, or other adult must supervise the work at all times.

- ▶ The team working on the project must communicate clearly, especially when turning on the gas and lighting the tube.

- ▶ Wear gloves, long-sleeve clothes, and eye protection.

- ▶ And always, always have a properly charged and checked fire extinguisher handy.

BUILDING YOUR RUBENS' TUBE

STEP 1: Take a piece of 5 foot long, 3 inch diameter ventilation duct, and cut it down to 3 feet long. The tubing is often sold as a long, C-shaped sheet of curved steel or aluminum with connectors stamped into the long edges that zip or snap together. You can go bigger, but you will go through more gas quicker when you fill it up and run the experiment. Be sure to cut the tube to length with metal snips before assembling it.

STEP 2: Use duct tape to cover the seam along the tube, and the sharp ends of the tube. This will help hold it together and protect the diaphragm (Glad Press'n Seal) from damage. Since leaking gas not only poses a fire hazard but will also reduce internal pressure and limit the height of your flame, doing everything you can to seal the tube is important.

STEP 3: Use a measuring tape or yardstick placed along the side of the tube opposite the seam and mark a dot every half inch along the entire length, leaving about 4 inches on each end (should be about 28 dots). Do not drill too close to the ends, since fire that close could cause the duct tape to melt or catch fire (a bad thing).

STEP 4: Using a drill press (much easier if you have one) or a hand drill, with the help of your mad scientist partner, drill a $1/16$-inch hole at every mark down the tube. If the bit wants to wander (especially with a hand drill), use a nail and hammer to tap a dimple at each mark first.

STEP 5: Because we don't want gas to escape anywhere except the holes, we'll need to seal up the ends of the tube. Any flexible and moderately heat-resistant material will work, like plastic wrap, a latex sheet, or even a large enough balloon stretched taut over the

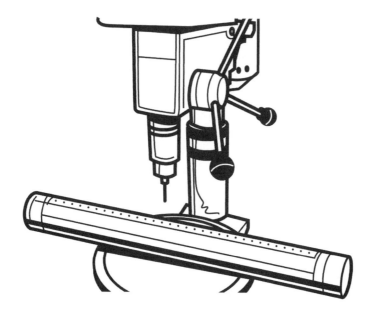

end. But for ease-of-use, Glad Press'n Seal (or the similar brand available in your area) is hard to beat. Stretch the material as you place it over each end and try to eliminate most wrinkles. Once you're satisfied with the fit, tape the sides down with duct tape.

STEP 6: For a flammable gas supply, we'll go with an inexpensive LP (propane) torch kit and cut off the nozzle on the end of the brass pipe so we can get a decent gas flow. We will need all the pressure we can get out of the can to fill the tube and provide a decent flame height. It is possible to go further and rig something up using a barbecue propane tank, but I'll leave that for the true mad scientists out there.

STEP 7: We're going to mount the tube onto the end of the propane tank so the gas will feed directly into the tube, and the tank will support the tube in the middle. You'll have to gauge the hole you drill based on the torch you use, but in our case, we drilled a $\frac{3}{4}$-inch hole that then had to be reamed out a little. Make sure the hole is next to, but not right on, the bottom seam.

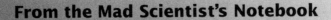

From the Mad Scientist's Notebook

While the tube will be supported in the middle by the torch, and is overall not very heavy, we do not want to rely on balance to keep us from accidently knocking the construct over while it's lit. You'll want to put heat- and fire-resistant supports under either end of the tube as well, and hopefully use some kind of clamping system to keep it from moving at all. What you have available in your laboratory is yours to explore, but always think about safety first!

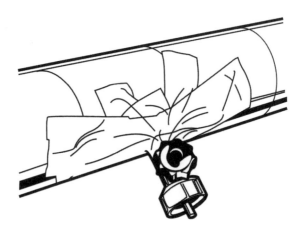

STEP 8: Tape the valve into place in the tube's hole. (Reminder: The valve does not need to be connected to the propane tank at this point—they are separate pieces, and it'll be easier to work with them unattached at this point.) Try to keep it neat to avoid leaks. Lay down enough tape to support the tube on top of the nozzle, and seal every possible hole (except the holes on top).

STEP 9: Connect the gas bottle to the valve and find a solid flat surface to set it on. The area should be clear of flammable materials and reasonably ventilated, but without wind or significant drafts. Prop up the ends and secure everything so it will not teeter or fall over if shaken or nudged.

STEP 10: Turn the gas on full at first. With one hand, spark a lighter near the center of the tube. Once it catches and lights, slowly turn the gas down. When all the holes are lit, adjust the gas so the flames are about $\frac{1}{2}$ to 1 inch high (sort of like lighting a gas barbecue).

Okay, your Rubens' tube is up and running. Now it's time to put on the show!

EXPERIMENTING WITH A RUBENS' TUBE

Start with the 440 hz sine wave sound (you can find a copy of the computer sound file here: http://en.wikipedia.org/wiki/File:Sine _wave_440.ogg) and play it through a speaker using your computer. The 440 hz tone represents a specific musical note, the A-above-high-C, and it considered the international standard for tuning instruments. It also has a very specific, clean visual wave form (invisible in the air), which you'll see with this experiment. Hold the

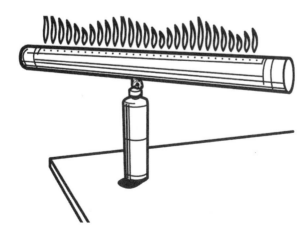

speaker up near one diaphragm and see how the fire responds. Adjust the gas so the flame does not go out, but isn't so out of control that you can't see the wave form.

Try different volumes to see if they change the shape of the wave. Find a different frequency or a sweeping frequency to see how they affect the shape and size of the wave.

Try different types of music. Just be careful with any heavy bass notes, as it could extinguish the flames. If that happens, just relight and try with a lower volume. And remember, the tube will get very warm, so never touch it with bare hands. When you're ready to turn it off, simply close the valve on the propane tank and the flames will die out. Open a window and/or turn on a fan to blow out any lingering gas.

This should be a visually exciting experiment. If you do it as a demonstration for a science fair, you may not be allowed to fire up the Rubens' tube at school, so instead you may want to take a video (with audio) of it working, with a nice black background to show off the flames.

Playing with Plasma

Idea by Kathy Ceceri

Most people think substance is more important than style, but for aspiring mad scientists, it's important to find a balance.

To look the part of a really cool mad scientist, one needs style; a certain joie de vivre! Black lab coats, steampunk goggles, and a fab retro-modern secret lair are all great accessories. But what's something even cooler you can create to pump up the atmosphere in your lab? Lightning bolts! Made with plasma! Now we're talking.

We all need a little more plasma—what one might call the most electric state of matter—in our lives. When I renovate my mad scientist lab next year, I'm thinking about installing a plasma chamber below the glass floor (you know, to replace the sharks-with-frickin'-lasers tank), so that every morning when I walk barefoot across the room to check on my experiments, you guessed it! Plasma will follow my steps. I'll be able to say I WALK ON PLASMA!

Okay, I may be getting a bit ahead of myself there. So first, why don't we just play with some plasma and get a better feeling for why we need more of this awesome state of matter in our lives?

EXPERIMENT	PLAYING WITH PLASMA
CONCEPT	Use inexpensive plasma globes with a number of other electronic components and explore the electromagnetic properties.
COST	$ $
DIFFICULTY	⚙ ⚙
DURATION	☼
DEMONSTRATION OR EXPERIMENT	Best as a demonstration.
TOOLS & MATERIALS	• Plasma globe ($10–$20 from novelty shops, museum gift shops, online) • Chair • Small fluorescent tube (may burn out a fluorescent compact bulb) • Small neon bulb (RadioShack has them cheap) • Xenon tube from old disposable flash camera • LCD screen (something expendable, like a dollar store calculator) • 2 pennies

This experiment (or series of experiments) works best as a demonstration of how plasma works, and how electrical current can move through gas, glass, metal, and even our bodies. It will work really well in a room where you can dim the lights to better see plasma dancing inside (and sometimes outside!) the sphere, as well as how it affects the components you use to interact with it.

First, set up the plasma globe where it is easy to reach without tripping on the electrical cord. Turn it on and try touching the glass. The tendrils of lightning will be drawn to whatever place(s) you touch. This is because your body acts as a conductor for the electrical charge moving from the generator (a small Tesla coil, more on this in the box on page 190) at the middle, through the gas, and to the outer glass shell. When you touch the glass, you're providing a new route for the charge to move through (like opening one spillway from a dam). But don't fear that you'll be electrocuted!

Most of the charge is blocked by the glass, and what passes through you is not dangerous. Try sitting in a chair and pulling your feet off the floor while touching the globe. Does it make a difference if you are "grounded" or not? It probably does. Being grounded means your feet (or some part of your body) is in contact with the ground, or the floor of your house, or whatever reasonably conductive surface you're standing on, so that the charge passing into you can pass right on out again. (Hint: Standing on a rubber bath mat will ensure that you're not grounded—the rubber keeps the current from passing through your feet into the ground.)

STEP 1: Set up the plasma globe on a table or other work surface. That's about it. Oh, make sure it's plugged in. And make sure you don't trip on the cord, or anything else. You might want to have a fire extinguisher handy, "just in case." But we're going to try and be very careful here, right?

STEP 2: Turn on the plasma globe and just touch it with your fingers. See how the tendrils of electricity flow to your fingers. There's actually electricity passing through the glass, into your fingers, through your body, and into the floor. If you sit without your feet touching the floor, the effect will be less because you're no longer grounded.

STEP 3: Now we'll test the electrical plasma field with a fluorescent tube.

As we've discussed, the charge from the plasma ball "leaks out" as electrons pass through the glass to the air, or to our fingers, through your body, and into the ground. It's also possible for the charge to pass into other electrical devices.

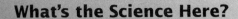

What's the Science Here?

If you made it through junior high (or even elementary) school science, you're probably familiar with the three states of matter we interact with the most every day: solids, liquids, and gases. There is a fourth, known as plasma, and while you may not have made the connection, you see it every single day. Plasma is a gas-like field of charged atomic particles—negative electrons and positive ions (atoms that have lost some of their electrons, and so have an excess of positrons). As they dash about, these particles generate electricity and magnetic fields. Plasma requires low pressure and extremely high temperatures. On Earth, plasma only occurs naturally in the form of lightning, polar auroras, and extremely hot flames. However, plasma is actually the most common state of matter in the universe, since it makes up stars (you know, like our sun), other celestial bodies, and the space in between.

Plasma was first identified in 1879 by Sir William Crookes, who called it *radiant matter*. It can be created artificially by running an alternating electric current through certain types of gas in vacuum tubes (you might possibly be familiar with one incarnation of this: neon signs). The current knocks electrons off the atoms of the gas, generating the super-charged material. Plasma globes like we're using here are Tesla coils (a type of transformer that takes standard electrical current and turns it into high-voltage, low current, high-frequency electricity that likes to jump rather spectacularly to nearby grounded conductors) inside a low-pressure glass vessel filled with gas. When the Tesla coil is fired up, it creates the plasma field of electrically charged particles, which look like small tendrils of lightning.

From the Mad Scientist's Notebook

Depending upon your or your kid's age, you may be more familiar with incandescent bulbs. In those, light is generated by passing electricity through a filament with a very high resistance. The filament heats up to such a temperature that some of heat energy becomes visible light, providing illumination.

But some kinds of light bulbs, like fluorescent tubes, light up when an electric charge excites gas molecules. They are, basically, another kind of plasma globe (they just don't have their own internal transformer—that's part of the lamp itself).

Fluorescent bulbs use molecules of mercury vapor, excited by the energy of the charged particles bombarding them from the plasma field generated by the electricity passing through. Electrons in the mercury atoms make a quantum jump to a higher energy level (or shell) around the atom's nucleus. When they return to their previous energy level, the extra energy is given off in the form of light.

If you pass a fluorescent tube back and forth near a plasma ball, it'll start to *fluoresce*, or glow, in sections near the globe. The charge from the plasma globe is passing through the glass, through the air, and into the tube, and is creating enough plasma there to generate light from the mercury vapor inside.

STEP 4: For a bit of extra fun, sit in a chair and hold one metal end of the fluorescent tube. Have someone who is standing hold the other end. Touch the plasma globe with your hand. If all goes well, the tube will light up. The human body is a good enough conductor that charge is passing from the globe, through you, through the fluorescent tube (lighting up the mercury vapor), out through your friend, and into the ground. You've created an electrical circuit out of yourself!

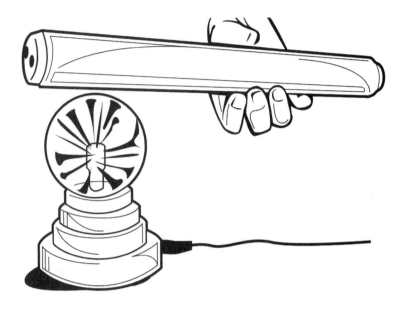

STEP 5: Now that we know what's going on (namely that our plasma ball is leaking electricity), we can play around a bit with other objects that might show an effect. Here are a few to hunt down and try:

- **Neon bulb:** You can find these at RadioShack or your local electronics store. They may come with a red plastic cover, but this is easy to remove, revealing the orange neon tube within.

- **Xenon tube:** You can take apart an old disposable camera and pull out the flash tube. Hold it so it's pretty much touching the globe.

- **LCD display:** Take apart a cheap calculator, disconnect the battery, and remove the LCD screen, circuit board, and printed plastic circuit sheet in one piece. Hold either the metal clips attached to the LCD screen or the plastic circuit sheet anywhere near the plasma globe. Touching the circuit sheet to the globe may create interesting effects (but don't leave it touching too long, or you may get a shocking result). Try testing the area of effect of the plasma field by observing the behavior of the LCD screen as you start out close to the globe, and slowly put it away. How far out does the field extend?

▶ **LEDs:** Try some cheap LEDs from RadioShack and see how easily they light up from the ambient electricity.

What Else Can I Do?

Caution—both these variations involved actually generating sparks outside the plasma globe. Sparks can burn, and they can even catch things on fire. Be careful, and be prepared (hence the earlier suggestion about the fire extinguisher)!

1. **Turn off the globe.** Place a penny on top of the globe so it won't slide off. Turn on the globe. Take another penny and bring it close to the first penny without touching it. Small sparks should appear. If you try using your finger instead of a second penny, you will feel a little shock. But you can create sparks to your finger without feeling a shock by lifting your feet off the floor!

2. **Turn off the globe.** Put a larger coin, say a quarter, on the top of the globe. Cover it with a piece of paper. Turn on the globe. Holding another coin, a metal nail, or a piece of wire, approach the paper where it's covering the coin. Sparks should jump through, and likely start burning holes in the paper.

There are many more things you can try—as many as you have electronic parts and gadgets. But always be careful—you are playing with electricity, which is inherently dangerous.

In the end, this experiment shows off some very important basic rules about how electricity works. It's also a great visual demonstration—especially if you can involve (un)willing participants in the show.

And when you're done, the added bonus is a plasma light for your mad scientist's nightstand. Because plasma is cool, and even mad scientists may be afraid of the dark.

Electrolysis
Splitting Molecules for Power

Idea by Kathy Ceceri

Nuclear fission. Splitting the atom (well, a few very specific atoms). Power. Mad scientists love power. Because if we had power (the electrical kind to run our gadgets, and the social kind to get everyone to do what we want), we'd really be able to run the world so much better than anyone else.

People so often accuse mad scientists of wanting power just so we can tell everyone what to do. That we enjoy power for power's sake and enjoy pushing people around.

In truth, mad scientists are altruists, seeking the betterment of humanity through our own tireless, often unsung efforts. We know, deep in our heart of hearts, that we can run things better. We understand how things should work, and we know that with the gentle, loving guidance of a kindly father figure in the right direction, everyone will be much happier.

The bossing-around part is just a perk.

So, power—the energy type used to heat water and spin turbines to generate electricity—is very handy in the pursuit of mad scientific goals. How else to keep our death rays raying, our destructobots destructing, or the lights on in our "Happy Camps of Optionally Mandatory Re-education"? Being able to create such power is an important arrow of knowledge in our quiver of things-that-are-useful-for-world-domination. And while we certainly aspire to splitting the atom and building our own nuclear power moonbases, at this stage, we'll set our sights a little lower.

Like splitting molecules instead.

EXPERIMENT	ELECTROLYSIS: SPLITTING MOLECULES FOR POWER
CONCEPT	We will break molecules into other compounds using electricity.
COST	$
DIFFICULTY	⚙ ⚙
DURATION	☼
DEMONSTRATION OR EXPERIMENT	Demonstration or experiment
TOOLS & MATERIALS	• Table salt • Tap or distilled water • 9-volt battery • 9-volt snap connectors (optional) • 2 old metal spoons • Medium-size glass bowl or jar • 9-volt battery cradle (optional) • Electrical tape (optional)

The basic experiment here involves some good old-fashioned electro-chemistry. We'll be using electricity from a battery to instigate a square dance of atoms and electrons that will break water molecules and dissolved salt into hydrogen and some new bits. Of course, gaseous hydrogen is rather flammable (no Hindenburg jokes, please—it's too soon), so we'll get a bit of fire out of this if we want to.

And we always want to.

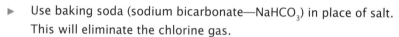

What's My Hypothesis?

As with any experiment, you need to hypothesize why something happens, and predict what might happen if you change something. For electrolysis, you can vary either the electrolyte or the material of the anode/cathode. The best thing to do is to try researching the different materials possible and, with a basic knowledge of chemistry, try to predict the result of using each one. Included here are some ideas:

▶ Use baking soda (sodium bicarbonate—$NaHCO_3$) in place of salt. This will eliminate the chlorine gas.

▶ Use graphite instead of spoons. One way is to take a pencil and sharpen it at both ends. See if that eliminates the green color change.

▶ Measure the change in pH in the water. Make a pH indicator (one method is to microwave some leaves of red cabbage, then use the water). Add some drops of vinegar to the water to start. Add the pH indicator. You should see the water change color from red to blue-green.

▶ Energy drinks like Gatorade contain electrolytes. Try using Gatorade for the liquid (or add some to water) to see if it is conductive.

▶ Test a number of substances to find their triggering amounts by slowly adding them to distilled water until the reaction starts (always use distilled water—mineral water already has potential electrolytes in it).

▶ You could even test mineral waters and sports drinks to see which ones have the strongest concentrations of electrolytes.

PERFORMING THE BASIC EXPERIMENT

STEP 1: First, we'll set up the bowl with our saltwater. The ideal ratio of salt to water for electrolysis is 10 percent salt to 90 percent water, calculated by weight (so the liquid measure on your measuring cups won't be quite as helpful). If you have a small kitchen scale, put the bowl on it, then zero out the weight (called taring) so that all you'll measure is the weight of what you add (if your scale doesn't support this, simply write down the weight of the empty bowl, and then keep track of the differential as you add materials). Add water until you've got the bowl about half-full and the weight you've added is easily divisible by 10 (so, for example, 10 ounces of water by weight is about 1 $\frac{1}{8}$ cups, or 300 grams of water is 300 milliliters—metric is so much simpler!). Take the weight of water you added, divide it by 10, and add that weight of salt by gradually dropping it into the water in the bowl on the scale and observing the differential.

STEP 2: Now for the electricity. We'll set up the battery and the leads very simply: Place the two spoons in the water, standing straight up with the bowls in the water. Be careful not to let the two spoons touch each other (this would cause a *short circuit*—the electricity from the battery would flow from one spoon to the other, directly, and completely miss the saltwater). Hold the ends of the two spoons to the battery's terminals, one spoon on each terminal. The spoon connected to the negative terminal is the cathode; the spoon connected to the positive terminal is the anode.

From the Mad Scientist's Notebook

Yes, this configuration is a little awkward, and might require an extra person to hold things together. You could also try picking up a 9-volt battery snap connector from your local electronics store (for example, RadioShack part #270-324). You know what I mean? The little black cap with the heads that snap onto the terminals of a 9-volt battery, with wire leads coming off. You've probably used them all your life. Well, you can buy them by themselves. In this case, you could connect the battery to the snap connector, and then tape the leads to the tops of the spoons instead, to make it all easier to handle.

STEP 3: Time for watching and recording the effects. Within a few seconds, you should see tiny bubbles coming off of the spoons. Hydrogen collects around the cathode and chlorine gas collects around the anode. You will notice what looks like smoke coming off the water—this is the chlorine gas (so, you know, don't inhale it).

After a minute or so, you should be able to see the water begin to turn murky yellow. After several minutes, the water starts to turn dark green. This is probably the result of the sodium hydroxide. As the water becomes more alkaline, it may also react with the metal of the spoons, producing the green color (like tarnished metal).

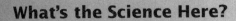

What's the Science Here?

We call water, colloquially, "good old H_2O." But why? Of course it's because that's its molecular formula, silly! Each molecule of water contains two hydrogen atoms and one oxygen atom. But here's a bit of crazy perspective for you: One 8-ounce cup of water—an amount we're supposed to drink multiples of each and every day—contains in the neighborhood of 7.91×10^{24} molecules of water. Let me write that out for you. 79,100,000,000,000,000,000,000,000 molecules of H_2O in every cup of water. Numbers that big are too much to wrap the mind around. Safe to say, molecules are very, very. . . . very small.

Anyway: The addition of energy in the form of electricity can make a water molecule split into its component parts (that's two Hs and an O—two hydrogen molecules and an oxygen). This process is called electrolysis. But pure water by itself is a poor conductor of electricity (yeah, we know you're never supposed to stand in a puddle near a downed power line, but that's an excessive case). What helps is if there's something else dissolved into the water that does conduct electricity. Such a substance is called an electrolyte. Basically it's any substance that has extra ions available to make it easy to conduct electricity (which is just the movement of charged particles—like electrons through metal wire). But in this case, the electrolyte is something we dissolve into the water to free up ions. Common electrolytes include acids, bases, and salts. Salts? Yup—so how about we make some saltwater?

Good old-fashioned table salt, known by its molecular name NaCl, makes for an excellent electrolyte when dissolved in water. The salt dissolves into its component parts, sodium (Na) and chlorine (Cl). But the sodium atom is missing one electron, giving it an overall positive charge. And the chlorine has an extra electron and so is negatively charged. These charged atoms, known as ions, are able to conduct an electric current through the water.

So, what actually happens? When you pass an electric current through the now very conductive saltwater, the water molecules (H_2O) break up. But, you don't end up with just hydrogen (H) and oxygen (O). The electrolytic salt you added recombines with the water molecules, leaving you with hydrogen, chlorine gas, and sodium hydroxide (otherwise known as lye). Here's the formula:

$$2 \; NaCl + 2 \; H_2O \rightarrow Cl_2 + H_2 + 2 \; NaOH$$

Let's take a look at the math there for a moment, because while it may seem like a foreign language to some people, it actually makes perfectly good sense once you've been let in on the secret.

Imagine each element is a differently colored 2 × 2 LEGO brick. The sodium (Na) is gray, the chlorine (Cl) is green, the hydrogen (H) is white, and the oxygen (O) is blue. At the start, we'd have four molecules, or clumps of bricks. We'd have two salt molecules, each represented by a gray brick stuck to a green brick. We'd also have two water molecules, each represented by two white bricks stuck to a blue brick. Can you picture that? Want to go get some bricks and set it up? I'll wait. . . .

All right, now we'll explain what happens in the reaction. Take all of the molecules apart into their component bricks. Don't put any away, and don't add any more. You should have two gray, two green, four white, and two blue bricks. Stick the two green bricks together. Stick two of the white bricks together. Stick one green, one blue, and one white brick together, and then repeat with the remaining bricks. You now have one molecule of chlorine gas (Cl_2—which is the element's preferred state at room temperature and air pressure), one molecule of hydrogen gas (H_2—same as the chlorine), and two molecules of sodium hydroxide, commonly referred to as "lye."

The point of this visualization was to demonstrate the fact that no material was added or lost. The equation is balanced, meaning both sides are equal. The things we started out with are the same as what we finished with, just jumbled around a little based on how they like to react to each other. That's basic chemistry!

Looking at the equation, you can figure a few things out as well. For example, the less salt you add, the more pure oxygen and hydrogen will be produced, and the less chlorine and lye will be formed. In addition, if you use metal lead to set up your electric circuit, some of the metal may react with the other molecules to create still more substances. Commercial electrolysis labs use other types of electrolytes, and non-reactive leads such as graphite, to make sure they only get the reactions and products they are looking for.

VERY IMPORTANT NOTE: Chlorine gas is toxic if inhaled, and lye (a corrosive alkaline) can burn your skin. However, the amounts produced by this experiment are very small. Even so, doing the experiment in a well-ventilated space and wearing eye protection and heavy rubber gloves is strongly recommended. Plus, it's a great excuse to don your full mad scientist outfit!

Here's a bit of fun to try: Add a drop of dishwashing detergent to the water. Soap bubbles containing hydrogen, oxygen, and chlorine gas will form. Hold a lit match to the bubbles—they will make a loud pop. You're lighting the hydrogen gas, which is very flammable (as always, be careful! In the absolutely impossible event that you get a 3-foot-diameter bubble of gases, just run away and call it a day!).

How can you beat that? An experiment that involves electricity, both explosive and poisonous gasses, and a corrosive compound should make any aspiring mad scientist giddy with the possibilities. It should also make you see how cool chemistry really is.

What's the Best Ammunition for Your Potato Gun?

Idea by Matt Forbeck

Ah, the potato gun. A cannon, really. Sort of like a training-cannon for the young, up-and-coming mad scientist. The basic device is an excellent tool for learning about combustion, pressure, and projectiles. Indeed, learning about projectiles and parabolic flight will come in quite handy once we perfect our inter-continental rail gun, so we can launch our robot-marauders anywhere we want from our stealth flotilla.

But I've said too much already.

EXPERIMENT	WHAT'S THE BEST AMMUNITION FOR YOUR POTATO GUN?
CONCEPT	There are many useful cannon fuels just lying around your house. We're going to build a cannon, and then test fuels for their efficacy.
COST	$ $ $
DIFFICULTY	⚙ ⚙ ⚙ ⚙ (Not technically difficult to build, but dangerous if done sloppily.)
DURATION	☼ ☼ ☼ ☼
DEMONSTRATION OR EXPERIMENT	The cannon can be used for demonstration or experiment, but only under adult supervision. It will likely never be allowed on the ground of a school.
TOOLS & MATERIALS	• 12-inch piece of 4-inch Schedule 40 PVC pipe • 48-inch piece of 2-inch Schedule 40 PVC pipe • 4-inch Schedule 40 PVC male adaptor with threads • 4-inch Schedule 40 PVC threaded plug • 4-inch to 2-inch Schedule 40 PVC reducer • Bottle of Schedule 40 PVC primer • Bottle of Schedule 40 PVC cement • Universal fit push-button grill igniter kit • 2 #10 2.5-inch machine screws • Drill and $5/32$-inch bit • Hacksaw • Tape measure • Marker • File • Wire stripper • *Propellant:* Good choices are Aqua Net or other aerosol hair spray, Right Guard or other aerosol deodorant, Lysol, Pam Cooking Spray, or Cutter Insect Repellent • Bag of fresh potatoes • Broomstick * **CAUTION:** Please note that we specified Schedule 40 PVC pipe for every piece. This is important for safety. Don't cheap-out and buy less sturdy materials.

SAFETY IS IMPORTANT!

Let me say that again.

SAFETY IS IMPORTANT.

Yes, we're going to build a cannon, and then experiment with propellants. If that goes just past your security zone, I'll totally understand, and suggest you head on to the next project. If not, let's just take a few moments to again talk about safety, because while building a cannon and shooting things out of it is a great deal of fun, it also has an element of danger. It's the kind of danger that can be minimized to safer-than-crossing-the-street levels of dangerousness by following instructions and a few specific safety protocols, but if said protocols are not followed, there is the chance for harm.

Follow the instructions. Use the materials listed above. High-strength Schedule 40 PVC is important because it can hold up to the explosions. You must work as a team with at least one adult in charge. You must follow key procedures, like never, ever standing in front of the cannon, or pointing the cannon at anyone, even if you think it isn't loaded. The adult parent/guardian is always in charge, and it is always a parent/guardian's prerogative to decide at what age their child is ready to participate in such a project.

Only use the cannon in a wide-open area with no people wandering around closer than at least 5 times the estimated range of the cannon. Every time you're going to launch, you must yell "Fire in the hole!" as loud as you can, twice in every direction. Indeed, actually using this cannon may be a challenge for people in an urban area, and you must consult your local codes to see if using such a device is even legal in your town.

Okay, onto the building.

HOW TO BUILD A POTATO CANNON

STEP 1: If you're lucky, some hardware stores will cut your PVC to your specified length, so go in with the aforementioned cut list and you might save yourself a step. Otherwise, use a tape measure, marker, and hacksaw to cut your PVC piping to fit the given lengths.

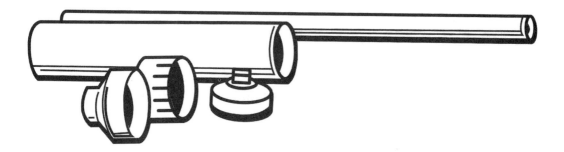

STEP 2: Use the PVC primer and cement per the printed directions to glue the threaded adaptor to one end of the 4-inch pipe, and glue the reducer to the other end. Use plenty of glue—we don't want anything popping apart at a critical moment because we didn't use enough glue! (Be sure you don't glue the threaded plug into the threaded adaptor. Indeed, you need to be able to take it out and put it back in fast when you fuel your cannon, so you should rub a little Vaseline on the threads instead.)

STEP 3: Glue the 2-inch pipe into the other end of the reducer.

STEP 4: Let the glue cure with the cannon open on both ends for at least 24 hours. You're going to be causing a controlled explosion in there, so it needs to be rock solid.

STEP 5: Take the grill igniter kit and attach the wires to it as if you were wiring it up for a side burner. Trim off the far ends of the wires and strip an inch of insulation off each of them.

From the Mad Scientist's Notebook

How long should I make the igniter leads?

A good question. This potato cannon is technically designed to be set off while being held, so you can cut the wires at just 6 inches if you like. However, you can further minimize risk by designing a system for a more remote ignition. If you give yourself about a 5-foot lead, and build a stand or tripod mount, you can then load the cannon and back away from the cannon to launch it more safely.

STEP 6: Drill a $^5/_{32}$-inch hole in the center of the 4-inch tube, then turn the tube 90 degrees and drill another hole lined up with the first. Drive the machine screws into these holes so that their tips are about $^1/_4$ inches apart from each other. Looking down the interior of the tube (okay, let's just get it over with and call it the barrel), the screws should form a "V" in the center, but with the point of the "V" not quite touching.

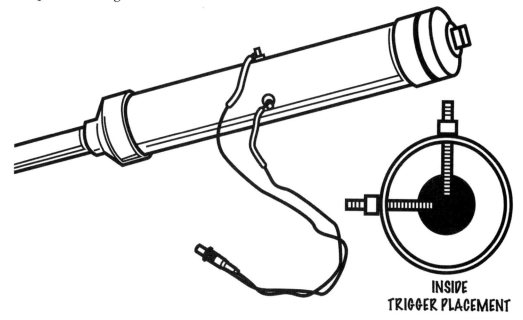

**INSIDE
TRIGGER PLACEMENT**

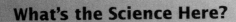

What's the Science Here?

This is cool, because we get to make a connection about a bit of history we mentioned in an earlier project. Remember the factoid about Pierre and Jacques Curie discovering the piezoelectric effect? Well, we're actually taking advantage of that effect right here in this project. You see, certain crystals will, under pressure, generate high-voltage electricity. Grill igniters work by making a spring-loaded hammer apply quick, hard pressure to a piece of crystal, generating a high-voltage charge. Positive goes down one wire, negative down the other, through the threaded screws in the pipe, and when each gets to the not-quite-touching tips, they jump, creating a spark. Just imagine what would happen if we had something combustible in there, under pressure?

STEP 7: Wrap one of the exposed wires from the igniter around the heads of each of the screws. Test the igniter by pushing its button, but keep your hands clear of the screws and exposed wires as you do. This should provide a small arc of electricity between the tips of the screws. If at first it doesn't work, adjust the screws in or out until you get a spark. Then cover them with electrical tape. You can also solder the wires to the screws, but electrical tape works just as well and insulates you from stray shocks, too.

STEP 8: Use a file to sharpen the end of the barrel. This makes it easier to cut the potato to the perfect size by just shoving it onto the barrel.

STEP 9: Decorate your cannon appropriately. Optional (as suggested above): Build a stand to mount your cannon on for remote-ish launching and ease of parabolic shots. Said stand must be able to handle the recoil of the launch and allow easy access to the threaded coupler so you can easily and quickly open and close the propellant chamber.

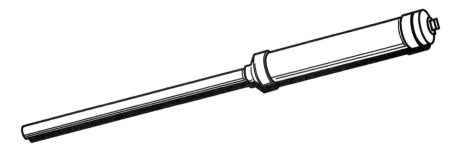

FIRING YOUR POTATO CANNON

STEP 1: Your big challenge is to find a wide-open area to fire the cannon, where you're not going to freak anyone out. You don't know exactly how far the potato might go, so be generous (as per the 5X suggestion above—try 500 yards safe distance). Also, wear gloves and safety glasses at all times.

STEP 2: Take a potato (obviously one that's larger than the barrel of the cannon) and shove it into the barrel, letting the barrel cut the excess away. This may take a little brute force, or might require some initial trimming to achieve. Make sure the potato fills the barrel and does not leave any gaps. Shove it down into the barrel all the way to the reducer with a stick or broom handle.

STEP 3: SPECIAL SAFETY PROCEDURE—Have your partner make a visual inspection of your target area to make sure it's clear, then have them yell out "Fire in the hole!" once in each of the four ordinal directions. Even as you prep the next step, have them keep an eye on the target area, just in case anyone or anything moves into it.

STEP 4: Now here's the high speed part of the launch, and you may want to practice without actually putting propellant in a few times, to get the hang of it.

Spray the propellant (Aqua Net hair spray is a great one to begin with) into the open back end of the cannon and screw the plug into it fast. Try spraying from a half-second to two seconds. What's key here isn't just filling the chamber with the propellant, but filling it with a mix of propellant and air. You need oxygen for combustion, right? So when you spray, don't coat the sides with the propellant, but rather try to get a cloud of it inside along with some air. Then quickly screw the plug in hand-tight.

STEP 5: Point the cannon's barrel away from any people or anything else you might harm with it.

CAUTION: NEVER point a potato cannon at a person, and NEVER fire it next to your head or between your legs for fear of the cannon fracturing or exploding. Wear gloves and appropriate eye protection. Keep children well away from it, and NEVER allow anyone directly behind it, either. Exercise caution at all times, for your kids' sake and your own.

STEP 6: If you're holding the cannon, brace yourself and push the igniter. Otherwise, step back from your mounted cannon, grab the igniter, grin at your partner, and push the button.

STEP 7: With a *whoosh* and a bit of a kick, you should get to watch your potato fly.

STEP 8: Potato cannon hygiene is important. Because un-burned propellant can build up in the combustion chamber, you need to clean the chamber out after every launch. Wet-wipes and/or paper towels can perform this task just fine. And while you may win the Battle of Agincourt with your new artillery, never leave a mess. When you're done launching, clean up all the remnants of your launches.

EXPERIMENTS FOR YOUR POTATO CANNON

Now that you've built your cannon, you have a variety of experiments you can run with it, mostly to see how far you can launch a potato. The easiest variable to work with is the propellant. Never use volatile, hard-to-handle fuels like, for example, gasoline, black powder, or lighter fluid. For safety's sake, it's best to stick with common aerosol sprays of one kind or another. These have flammable elements that work well with a potato cannon but aren't usually so volatile as to cause the cannon to explode.

Try, for instance:

▶ Other aerosol hair sprays (you can even research the ingredients to see what concentrations of certain chemicals might work better or worse)

▶ Right Guard or other aerosol deodorant

▶ Gumout Carb & Brake Cleaner

▶ Lysol

▶ Pam Cooking Spray

▶ Cutter Insect Repellent

When fueling the cannon, use a roughly equal amount of spray, either one or two seconds' worth. Just do your best to keep it consistent.

When doing launch experiments with your cannon, consistency with respect to the angle of fire will be key, and it's an even better idea to design and build a launch mount for your cannon so each time you launch it with a different propellant, you launch it at the same angle. Then you can roughly pace off each distance, or if you have access to a very long construction tape measure, use that.

Alternatively, if you have access to a speed gun, you can measure the muzzle velocity of each discharge instead.

Figure out the cost per ounce of each propellant. Then you can calculate roughly how far you can fire the cannon for each ounce's worth of fluid, aka how much bang you get for your buck (literally!).

OTHER POTATO CANNON AMMUNITION

You can experiment with different types of ammunition for your potato cannon as well. Try, for example:

▶ Sweet potatoes

▶ Apples

▶ Corn on the cob

▶ Red potatoes

▶ Lemons

Further Reading

If you get stuck with your construction or are looking for more advice, consult the following sites:

▶ http://www.spudtech.com/
▶ http://www.spudgun.com/
▶ http://potato_cannon.webs.com/
▶ http://aaroncake.net/spuds/index.asp

And so now you can start dreaming of your giant potato gun, with a 12-foot diameter barrel. It'll take seventeen cans of hair spray to prime, and you'll need to genetically engineer 280-pound potatoes, but the ability to shell a town three counties away with spuds is really worth it.

Deadly Dust
A Ticking Bomb?

Idea by Matt Morgan

Did you know there's a bomb waiting to explode in your pantry? Heck, more than one. Let me explain.

Death from the dust; sounds like a good old pulp thriller, doesn't it? Well, since the tragic 1878 explosion of the Washburn flourmill in Minneapolis, Minnesota, it's also recognized as a serious danger in the baking goods industry. Yes, the flour (or baking powder, and maybe the powdered sugar) in your kitchen cupboard is a little bit of TNT just waiting for you to light the match.

And you, you're an aspiring mad scientist. You're going to light that match—in controlled scientific conditions that allow you to study the reaction, of course!

EXPERIMENT	DEADLY DUST: A TICKING BOMB?
CONCEPT	Demonstration of how increased surface area can greatly speed the reaction rate of combustion (a chemical reaction).
COST	$
DIFFICULTY	✿ ✿ — ✿ ✿ ✿
DURATION	☼
DEMONSTRATION OR EXPERIMENT	Demonstration or Experiment
TOOLS & MATERIALS	• No. 5 coffee can (or similar). Half gallon sized, approximately 59 fluid ounces • Bag of flour • Candle • Clay or putty (optional) • Nail or tack (optional) • Box of matches or barbecue lighter • Turkey baster • Measuring spoon (tsp.) • Small bowls • Electric drill • Goggles • Damp hand towel

Everything burns. Scary thought, isn't it? Of course it takes a LOT of energy to ignite some stuff (the aluminum case of your laptop isn't likely to burst into flames in direct sunlight, for example). On the other hand, materials like wood, gasoline, and oil are happy to burst into flame at the touch of a random spark. The thing is, through our collective wisdom we think we know what's dangerous and what's not, and so we treat the flame-y stuff with due respect.

But do we really? How many of us think about the big bag of flour on our pantry shelf and realize we should have a fire extinguisher handy nearby?

Actually, that's an exaggeration. But it's not hard to make flour very flammable, and that makes for a great experiment. It's all about the dust, so we're going to make a combustion chamber and light some of the powders we have lying around the kitchen.

MAKING THE COMBUSTION CHAMBER

STEP 1: You need to drill a hole in the side of the can wide enough for the tip of the turkey baster to be inserted, probably $\frac{3}{4}$ inch to 1 inch in diameter, and 1 inch to 2 inches up from the base of the can. The baster does not need to extend more than 1 inch into the can, and we don't want the hole so high that when we blow air into the can, it snuffs out the candle. And your candle should be short enough to leave 3 to 4 inches of clearance with the top of the can, so you may need to do a little math (or a little candle-cutting) to make everything fit correctly.

STEP 2: Place a candle in the center of the can. You might want to use some clay or putty to make it stick to the base. Or you could hammer a nail or tack through from the underside of the base so the candle can sit on the point and stay well anchored. We don't want our puffs of air (or the bumping that comes with it) to knock the candle over.

STEP 3: FOR SAFETY—get a hand towel, run it under the tap until it's wet, wring it out, and have it nearby. Also, goggles are a good idea (and an excellent mad scientist accessory). I'll point out here that this is one of those projects where you'll want to make sure your budding mad scientist understands the safety procedures very clearly. You are creating an explosion. There will be fire. You both want to keep your eyebrows and hair intact. Be deliberate and cautious. While this is a very fun experiment to run, never play around in the lab. You and your kids may want to become mad scientists, but you'd really rather avoid the ever-so-stereotypical lost-limbs or hideous scarring that the sloppy mad scientists have, wouldn't you?

MAKING THINGS GO WHOOSH! AND BOOM!

STEP 4: First, you're going to want to do the "before" experiment, which shows why we're not so nervous about the bag of flour in the pantry. Pour several teaspoons of flour into a separate, heat-safe bowl.

STEP 5: Light a match and drop it into the bowl of flour. WHOOS— err, maybe not. While the flour may smolder, you're really not going to get much effect. There's a reason for that we'll explain on page 200. Use a spoon to fish the match and burnt flour out of the bowl, then rinse them in the sink to make sure they're no longer on fire, and dispose.

STEP 6: Now, place several teaspoons of flour in the can piled into a mound just adjacent to the hole you made earlier in the side. You want to be able to stick the tip of the turkey baster through the hole and bury it in the flour.

If you need to add to the pile, do so in small amounts. More flour in the can will result in a larger explosion, so be conservative to start.

STEP 7: With everything set (and all safety measures in place), you can light the candle. Be careful while you do so as not to introduce any puffs of air into the can (DON'T SNEEZE! IT WOULD BE TOO MUCH LIKE A BAD SITCOM MOMENT!), because that might start the explosion before you're ready for it (usually a bad thing).

STEP 8: Insert the baster through the hole in the side of the can and bury the tip into the mound of flour. Apply a steady puff of air from the baster by giving it a quick, sharp squeeze (but don't jar the can). The flour should poof into a cloud, disperse through the air in the can, and ignite in the flame, generating a quite satisfying *whoosh!*

TIPS, TRICKS, AND TROUBLESHOOTING

If you had trouble calibrating your candle height in conjunction with your baster placement, you might find that a quick puff of air will put the flame out before the flour disperses and ignites. If this happens, try substituting cornstarch for the flour. Cornstarch is lighter, and will more easily poof into a dust cloud.

If you want to further increase the chance of a dust explosion, use multiple candles of varying heights, or candles with multiple wicks. Obviously, the more fire at the start, the easier it will be to get the reaction going.

For even more fun (and a louder noise), place the lid on the can just prior to blowing the air in to make the dust cloud. (DO THIS IN THE GARAGE OR OUTSIDE AND MAKE SURE YOUR PETS

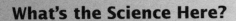

What's the Science Here?

The flame occurs from the even application of heat to the flour so that the powder can get hot enough to ignite. In the bowl, with just a match tossed on top, individual grains couldn't absorb enough heat to start burning, cause adjacent grains to burn, and start a true conflagration where the fire keeps seeking out fuel to continue.

But when we dispersed the flour with a puff of air and turned it into a cloud of dust, individual particles could be completely surrounded by hot air and flame, allowing them to far more easily reach the temperature required for ignition. Once particles started to ignite, the energy they expelled provided enough additional heat to ignite neighboring particles. The resulting chain reaction in this case is a dust explosion.

AREN'T NEARBY.) If you puff the dust before the candle can use up all the air inside (or melt a hole in the lid), the resulting explosion should be enough to literally blow the lid off the can.

Again, safety first—never stand so that any part of your body is over the can when you try this. And you might want earplugs.

In the end, this is a fun and dramatic demonstration as to how easy it is to turn something as innocuous as baking flour into a real explosive. There's not a lot to hypothesize and experiment with, unless you explore different types of dust materials (but kids should really check with their parents first, to make sure they aren't using something *really* dangerous). But the lessons about combustion are very valuable, and if you're lucky, this will be the thing that gets your little mad scientist thinking about what *else* they can burn or blow up (under the appropriate, safe, and well-monitored conditions, of course!).

Emission Spectroscopy
Identification with Flames

Idea by Kathy Ceceri

F ire. The spark of human progress. The destroyer *and* the creator of stuff all rolled into one. Or, to quote the scholars Beavis and Butt-Head, "Fire!, heh-heh, Fire!"

Humans have a natural fascination with fire, and mad scientists doubly so. Fire is the realization of the Sword of Damocles metaphor: the thing so useful, and so dangerous.

Which is why we're going to have some fun with it. But not just fun—scientific fun. Because while flame can destroy, it can also unlock secrets. And if there's anything a mad scientist loves almost as much as wanton destruction, it's learning other people's secrets!

EXPERIMENT	EMISSION SPECTROSCOPY: IDENTIFICATION WITH FLAMES
CONCEPT	Fire can unlock the secrets of various materials by giving off signature colors when burning them. We'll demonstrate this effect by burning a variety of materials and observing the flame colors that result from the combustion.
COST	$ $
DIFFICULTY	⚙ ⚙ ⚙
DURATION	☼
DEMONSTRATION OR EXPERIMENT	Demonstration
TOOLS & MATERIALS	• Candle or propane torch • Cotton swabs or wooden splints • Set of various metal salts [Strontium Chloride, Sodium Carbonate, Cupric Sulfate, Potassium Chloride, Cupric Chloride]. (See below for how to procure these) • Bowl of water per metal salt • Insulated tongs • Fire extinguisher

As we've explored throughout the projects in this section, everything can burn, given the right circumstances—even things like the metals that we take for granted in our everyday lives as very strong materials.

And while we've alluded to the idea of how a flame's color can tell us things, we haven't talked about what that means. The interesting truth is that when metals burn, they give off colors—different colors for different metals! It all started in 1835 when English scientist and inventor Charles Wheatstone was the first to scientifically demonstrate that different metals each had a uniquely colored spark. From there, the sky (or at least the lab) was the limit, and thus was born spectroscopy, the identification of materials by the color they give off when burned.

So, what to make of all this? Well, it's always fun to do a bit of

sleuthing, and learning how to suss-out a material just by burning it sounds like fun! Really, this is a better demonstration project than experiment.

Metal salts can easily be purchased as part of an inexpensive kit from any reputable science shop. There are many online sites that offer a set of five or more salts for approximately $15, but several of these flame colors can be achieved through the burning of common household products (see the list in Step 4).

EXPERIMENT (AKA "BURN, BABY, BURN!")

STEP 1: Prepare the chemical solutions by mixing metal salts with water in a non-metallic container. The solution should be a mixture of 1 ounce of metal salt per 1 cup of water. Perform this outside, in a workshop space, or on a durable/cleanable surface as some solutions can stain.

STEP 2: Select an object that will absorb the solutions and be burnt. This could be wooden splints, or cotton swabs, but for wooden splints, you'll need to soak each splint in its assigned solution for at least 24 hours. Whereas cotton swabs can be soaked in much less time, approximately one minute. The point is to saturate the burning media with the solution.

STEP 3: Minding all the safety provisions listed in the Introduction, burn the soaked materials one by one and observe the colors produced in the flames. You can use a candle, propane torch, or perhaps a caramelizing torch available in many kitchen stores. You can write down your observations for each, or even better, take video of each burn so you can re-examine and even freeze-frame the flames to get a good sense of the color. You may want to have a neutral, white background behind the flames so the colors are easier to see.

Note: *If flame colors are being masked by a predominantly yellow flame, the yellow light can be filtered out by viewing through blue-tinted glass.*

STEP 4: The results of the demonstration can be interpreted by comparing the flame colors to this rough guide of the spectrum and its corresponding chemicals:

Red—Strontium Chloride

Yellow—Sodium Carbonate

Green—Cupric Sulfate (tree root killer for plumbers)

Light Green—Borax (laundry and cleaning solution)

Violet—Potassium Chloride (water softener salt)

Pink—Lithium Chloride

Blue—Cupric Chloride

White—Magnesium Sulfate (Epsom salts)

Warning: While the preceding compounds are safe to burn, do not experiment with others not listed. Specifically, permanganates, nitrates, and chlorates will produce harmful byproducts if burned. You can always do more research to determine other compounds to try. If you're truly adventurous, pick up one of those "Chem C1000" chemistry sets, which will have far more materials and information to work with. It's one of the only good chemistry sets left on the market.

What's the Science Here?

During combustion, enough heat energy is transferred to burning materials to promote its electrons to a higher energy level, overcoming the electrostatic attraction that would normally keep such electrons in place. The change in position is short-lived, and as electrons fall from this excited state back into a stable state, their excess energy is cast off in the form of light of various wavelengths, depending upon the amount of energy being released.

The amount of energy required to boost electrons into higher states is unique to each element, defined by its atomic energy level. When this excess energy is cast off as light (in the form of a photon), the amount of energy will dictate what portion of the visible spectrum (color) this light will occupy.

This change purge of excess energy is referred to as an energy level transition, and the amount of energy in a photon is equal to its frequency multiplied by plank's constant. The inverse of this energy defines the photon's wavelength, which can be compared to the visual spectrum to determine color.

Comparing these results to previously observed values allows scientists to identify the material(s) that were burnt. This process is called emission spectroscopy.

If you want to present emission spectroscopy as a demonstration experiment, first make sure you'll be allowed to have an open flame where you're performing . . . uh, experimenting! We mean experimenting! You're going to do this like a magic trick, but using the magic of science!

Make up your solutions beforehand and put them in identical sealed containers for transportation. Label each container only on the bottom as to the material inside.

When you're ready to demonstrate, set the containers on a tray and ask an observer to switch them around, like hiding the peas under the proverbial nuts in the classic magic trick. Don't watch them do it. When they're done, mark each container with a number on the side.

Prepare your burn samples (you should probably use the cotton swabs, for the sake of time). One by one, burn the samples and write down the color and the number container that the sample came from.

When you're done, it's time for the flourish. If you memorized the color signatures of each of your materials beforehand, you can play that up and simply point to each jar and tell your astounded audience what the material inside is. Have your audience member check the name on the bottom of each jar to validate your findings. And with that, you're a magician of science!

APPENDICES
Appendix A

Experiment Index

EXPERIMENT	COST	DIFFICULTY	DURATION
EXPERIMENTS FOR MOONBASE ALPHA			
Extracting Your Own DNA	$	🦷🦷🦷	🔆
Space Agriculture	$	🦷🦷	🔆🔆🔆🔆
DIY Mind Control	$–$$	🦷🦷–🦷🦷🦷	🔆🔆–🔆🔆🔆
Growing Crystals for Power	$	🦷	🔆🔆🔆🔆
Biosphere Breakdown	$–$$	🦷🦷	🔆🔆🔆🔆
Can You Dodge a Laser?	$$–$$$$	🦷🦷🦷–🦷🦷🦷🦷	🔆🔆🔆🔆
Spaceship Design: Building Your Own Wind Tunnel	$–$$$$	🦷🦷🦷–🦷🦷🦷	🔆🔆🔆–🔆🔆🔆🔆
INSIDE THE MAD SCIENTIST'S KITCHEN			
Making Your Own Topsoil	$ $	🦷	🔆🔆 🔆🔆
Growing Tasty Sea Monsters	$	🦷	🔆🔆 🔆🔆
Exploring Fluid Dynamics	$–$$	🦷	🔆
Understanding Calories: Junk Food in Flames	$	🦷🦷🦷–🦷🦷🦷	🔆🔆
Thermodynamics: Keeping It Cool without Electricity	$	🦷	🔆🔆

EXPERIMENT	COST	DIFFICULTY	DURATION
APOCALYPSE SURVIVAL SCIENCE			
Building a MacGyver Radio	$	⚙⚙	☼☼
Making Top Secret Invisible Ink	$	⚙⚙	☼☼
Steam Power, Steampunk Style	$–$$	⚙⚙–⚙⚙⚙	☼–☼☼
Mastering Alchemy	$	⚙	☼☼
The Science of Siege Warfare	$–$$$	⚙⚙–⚙⚙⚙⚙	☼☼☼
Post-Apocalypse Particle Detector	$$	⚙⚙⚙	☼☼
Mapping Your Ecosphere	$	⚙	☼–☼☼☼☼
FUN WITH FIRE AND ELECTRICITY			
The Music of Fire—Building a Rubens' Tube	$$–$$$	⚙⚙⚙–⚙⚙⚙⚙	☼–☼☼
Playing with Plasma	$$	⚙⚙⚙	☼
Electrolysis: Splitting Molecules for Power	$	⚙⚙	☼
What's the Best Ammunition for Your Potato Gun?	$$$	⚙⚙⚙⚙	☼☼☼☼
Deadly Dust: A Ticking Bomb?	$	⚙⚙–⚙⚙⚙	☼
Emission Spectroscopy: Identification with Flames	$$	⚙⚙⚙	☼

Appendix B

Arduino Code for "Can You Dodge a Laser?"

[Note: The following code will also be hosted as a downloadable file in the forums on www.geekdadbook.com for easier input.]

```
const int ready_light = 2;
const int fire_light = 3;
const int lazzor = 4;
const int fire_switch = 7;
const int sensor_in = 0;
int seed_timer = 0;
long sensor_val = 0;


void setup(){
pinMode(fire_switch,INPUT);
pinMode(ready_light,OUTPUT);
pinMode(fire_light,OUTPUT);
pinMode(lazzor,OUTPUT);
Serial.begin(9600);
}

void loop (){
//check_alignment();
//read_sensor();
fire_the_lazzor();
```

```
}

long read_sensor(){
int base_line = 0;
int trip_point = 3 * base_line;
sensor_val = analogRead(sensor_in);
if (sensor_val > 500){
digitalWrite(ready_light,HIGH);
return sensor_val;
}
else{

digitalWrite(ready_light,LOW);
return 0;
}

}

void check_alignment (){
int count =0;
long sensor_read = 0;
for (int i=0; i < 4; i++){
digitalWrite(ready_light, HIGH);
delay(250);
digitalWrite(ready_light,LOW);
delay(250);
}
Serial.println("Calibrating Laser Please wait");
digitalWrite (lazzor,HIGH);
while (count < 10){
sensor_read = read_sensor();
Serial.println("Adjust Lazzor until Ready Light turns on");
if(sensor_read > 300){
Serial.print("Sensor Reading ");
Serial.println(sensor_read);
count++;
delay(100);
}
else{
count=0;
```

```
}
}
Serial.println("Lazzor is aligned . . . commence firing");
//return 1;
}



void fire_the_lazzor(){
long random_seed=0;
long delay_time=0;
static unsigned long startTime = 0;
static unsigned long reactionTime = 0;
digitalWrite(fire_light,LOW);
digitalWrite(lazzor,LOW);
Serial.println("Press The Fire Button");
Serial.println("When the Red Light goes on the lazzor will fire. . . . ");
random_seed=switchTime();
if (random_seed > 0){
delay_time = random(random_seed);
delay(delay_time);
digitalWrite(fire_light,HIGH);
digitalWrite(lazzor,HIGH);
//delay(500);
sensor_val = analogRead(sensor_in);
}
while(sensor_val < 500){
digitalWrite(lazzor,HIGH);
}
reactionTime = millis()-startTime;
Serial.print("Reactioj Time is . . . ");
Serial.println(reactionTime);
Serial.println("Hmmm took longer than I expected . . . looks like you
    got burned");
```

More Geeky Ideas
from Ken Denmead

United by the premise that to really understand science and how something works you must design and build it yourself or remake it better, Ken Denmead's *New York Times* bestselling series provides geeky families with hours of DIY fun.

GOTHAM BOOKS a member of Penguin Group (USA) | www.penguin.com